JET AGE
FLIGHT HELMETS
Aviation Headgear in the Modern Age

Alan R. Wise & Michael S. Breuninger

Schiffer Military History
Atglen, PA

ACKNOWLEDGEMENTS

The knowledge and assistance of many individuals, be it a collector, historian, veteran, or manufacturer was required to complete this book. Our utmost thanks to the following for their time and contributions: Scott Ferguson - United States Air Force Museum, L.P. Frieder, Jr., President - Gentex Corporation, Trevor Garrod, Dennis H. Gilliam, Al Kozuch - Gentex Corporation, N.A.S.A./Johnson Space Center, Lee F. Patterson, Anders Skotte, Leo Unruh - Gentex Corporation, Alexander Varlet.

Special thanks to: Max Ary and Steve Garner, Kansas Cosmosphere and Space Center, Thomas Blaine - Gentex Corporation, Thomas Asher, Kay L. Breuninger, Alexander 'Skip' Clawson - Gentex Corporation, Helmet Integrated Systems, Ltd., Steve Griffith, Lillian Kozloski - National Air & Space Museum/Smithsonian Institution, Douglas M. Palmer, J.W. Peklo, Brad Schoewe, David J. Shablak, Ruth W. Shablak, John Spencer, United States Naval Aviation Museum, Steve Wilson.

Book Design by Robert Biondi.

Copyright © 1996 by Alan R. Wise & Michael S. Breuninger.
Library of Congress Catalog Number: 96-67284.

All rights reserved. No part of this work may be reproduced or used in any forms or by any means — graphic, electronic or mechanical, including photocopying or information storage and retrieval systems — without written permission from the copyright holder.

Printed in China.
ISBN: 0-7643-0070-9

We are interested in hearing from authors with book ideas on related topics.

Published by Schiffer Publishing Ltd.
77 Lower Valley Road
Atglen, PA 19310
Please write for a free catalog.
This book may be purchased from the publisher.
Please include $2.95 postage.
Try your bookstore first.

CONTENTS

ACKNOWLEDGEMENTS 4
INTRODUCTION 6

Chapter 1 U.S.A.F./U.S. ARMY FLIGHT HELMETS 9
        HIGH ALTITUDE HELMETS 57

Chapter 2 U.S.N./U.S.M.C./U.S.C.G. FLIGHT HELMETS 87
        HIGH ALTITUDE HELMETS 122

Chapter 3 FOREIGN FLIGHT HELMETS 129
        CANADA 129
        CHINA 132
        FRANCE 142
        GERMANY 153
        POLAND 155
        RUSSIA/SOVIET UNION 156
        SWEDEN 177
        UNITED KINGDOM 188

Chapter 4 SPACE HELMETS 212
        RUSSIA/SOVIET UNION 212
        UNITED STATES 228

INTRODUCTION

Born in the waning days of World War II, the Jet Age coincided with the dawn of the nuclear era and the Cold War. The combination of these realities resulted in an unprecedented development of aircraft capable of supersonic-plus speeds, phenomenal altitudes, and undreamed of maneuvering dynamics. The advancement of pilot headgear closely paralleled this rapid progress and in fact played a crucial part.

While seemingly unnecessary within the safe confines of a Jet Age aircraft's pressurized, climate-controlled cockpit, headgear needs to reduce the powerful engine noise to a comfortable level and provide the pilot with a means to communicate vital information. Oxygen must be supplied, and intense sunlight must be filtered. To reduce strain on the pilot's neck during high-G maneuvers, overall headgear weight must be minimal and for long duration flights, it must be comfortable. So much for the easy requirements. Should the pilot ever face the distressing necessity to eject from a supersonic aircraft, at altitude, an inhuman world awaits with the simple pull of a handle. Instantly, he is subjected to a windblast capable of tearing his skin and eyes from his face, if he were not protected. The temperature could be thirty below zero Celsius, and his essential need for oxygen continues. When his parachute deploys at high speed, the snagging of lines or risers on his helmet could cause a broken neck or fouled canopy. When he lands in rough terrain, impact protection for the most vulnerable part of his body must be protected by his helmet. And to be found, especially in the sea, he might need the aid of a brightly-colored surface. In combat, he would desire the opposite. In practical terms, Jet Age headgear is not only a helmet, it is a complete system, and it is life.

To the high-altitude pilot, cosmonaut, or astronaut, the pressure helmet is part of a critical life-support system, a thin barrier between life and instant death.

Aesthetically the helmet represents the skill, dedication, and courage of the pilot. He is immediately recognized by his headgear. And perhaps nothing else so clearly represents the interface between Man and the most sophisticated of all man-made machines, the combat aircraft or spacecraft.

In creating this photographic chronicle of Jet Age aviation and space headgear, it is our intention to introduce the aviation enthusiast, collector, historian to an extraordinary history within a history. We did our very best to provide as much information and coverage as we could within the realm of a photographic record. Whenever possible, color photographs were used; when not, archival photographs were either substituted or used to compliment.

It was our considered decision to limit the headgear covered to only those examples which were considered either operational or used in major experimental programs such as the X-15 project. While a few exceptions are included due to their profound design influence, such as the X-20 "Dyna-Soar," or due to their historical significance, such as the Project Apollo A1C, to have included all experimental and unsuccessful prototypes would have been unmanageable. It was also impractical to cover the headgear of every air force in the world. The major aircraft producing countries, China, France, Russia, Sweden, United Kingdom and the United States produce most modern headgear, and other countries overwhelmingly use the source of aircraft as the source of headgear. For example, the Iraqi Air Force uses Russian headgear in their MiG-29's and Mi-8's and French in their F-1's. Japanese Defense Force and Israel Defense Force pilots use U.S. produced headgear of the same specifications as those used by USAF pilots. Examples produced in one country to very unique specifications for another are included in a limited manner. Any other omissions are unintentional and the authors would be interested in learning the details.

A U.S.A.F. pilot wearing his painted P-1 helmet in 1954. The 82nd Fighter Squadron flew F-94B and C aircraft. (Photo courtesy of L. Richard Hornbeck)

CHAPTER ONE

U.S.A.F./U.S. ARMY FLIGHT HELMETS
&
HIGH ALTITUDE HELMETS

P-1

The P-1 helmet, introduced in 1948 and made by General Textile Mills (renamed Gentex Corporation in 1958), was the first standard issue Air Force hard helmet. It was used during Korea in fighter aircraft and was also procured by NATO and other friendly countries. This and other early 'P' helmets were later retrofitted with external visors and upgraded avionics. The 'P' series helmet shells are constructed of a fiberglass cloth reinforced with epoxy resin. The MS22001 is similar to the A-13A oxygen mask of late World War II and was redesignated by the U.S.A.F. in 1950.

U.S.A.F P-1 helmet with B-8 goggles and MS22001 oxygen mask.

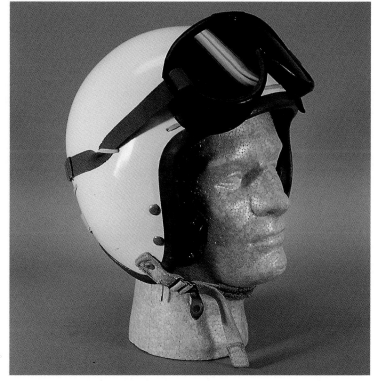

The chin (or throat) strap was a 1950 modification replacing the chin cup. These modified P-1's were redesignated as P-1A's.

MS22001 oxygen label.

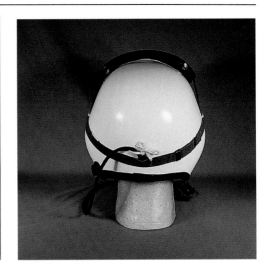

Early leather chin cup attached to the snaps on leather tab.

Method of securing the goggle strap. The MC-3A oxygen connector is also shown.

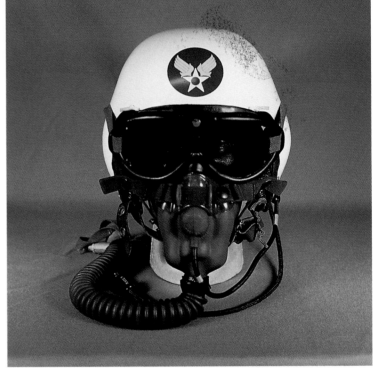

A-13

U.S.A.F. A-13 leather covered, reinforced helmet with G-2 sunglasses, MS22001 oxygen mask and MC-3A oxygen connection. The A-13 helmet was used in the mid 1950s in such bomber aircraft as the B-36 Peacemaker.

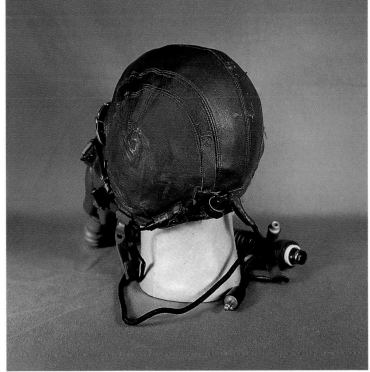

Helmet adjustment strap and external goggle retaining straps.

P-1A

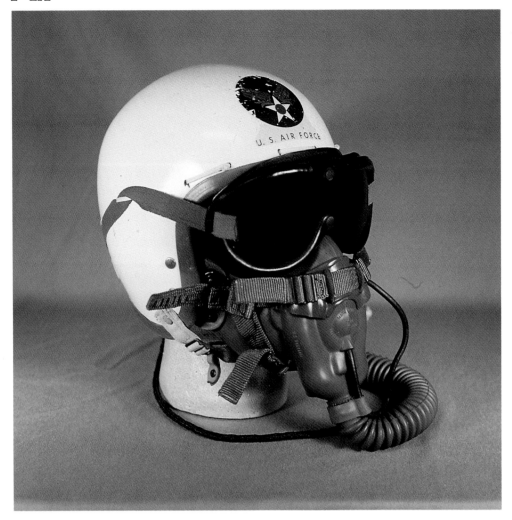

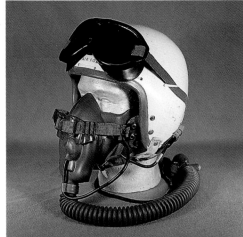

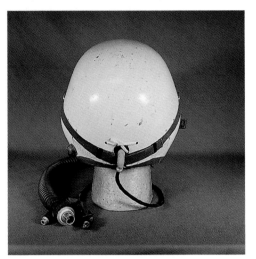

U.S.A.F. P-1A helmet with B-8 goggles and MS22001 oxygen mask. The P1-A was adopted in 1949 and was virtually identical to the P-1 helmet except for the addition of a nape strap. It was used during Korea in such fighter aircraft as the F-86 Sabre and declared obsolete in 1959.

The MC-3A oxygen connector is visible.

The majority of 'P' series helmets used leather tabs and snap fasteners to secure the oxygen mask.

CHAPTER 1: U.S.A.F./U.S. ARMY FLIGHT HELMETS & HIGH ALTITUDE HELMETS

P-2

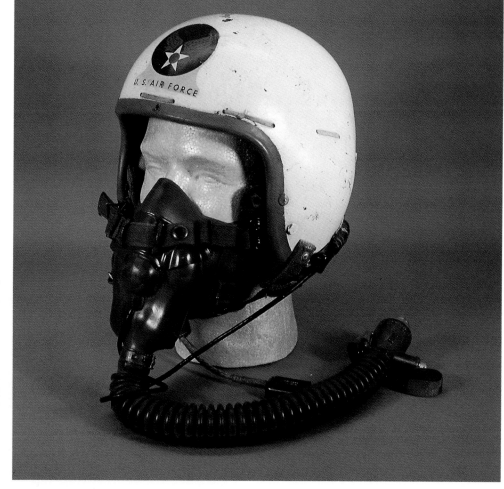

U.S.A.F. P-2 helmet with MS22001 oxygen mask.

U.S.A.F. P-2 helmet.

P-1B helmet retrofitted with an external visor. Note how the center locking track is attached directly over the Air Force decal.

P-3

Tinted visor in down and locked position (Note the spring tensioners).

U.S.A.F. P-3 helmet with external visor and MS22001 oxygen mask. The P-3, standardized in 1951, was used in fighter and bomber aircraft and declared obsolete in 1959. The external visor was added to the P-3 helmet in 1952 and did not use a center locking track.

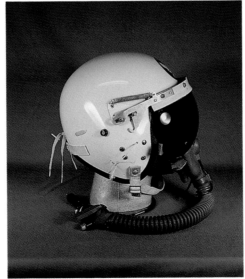
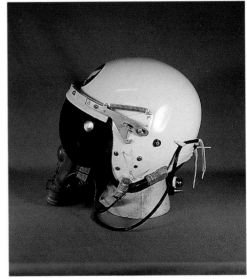

CHAPTER 1: U.S.A.F./U.S. ARMY FLIGHT HELMETS & HIGH ALTITUDE HELMETS

P-4

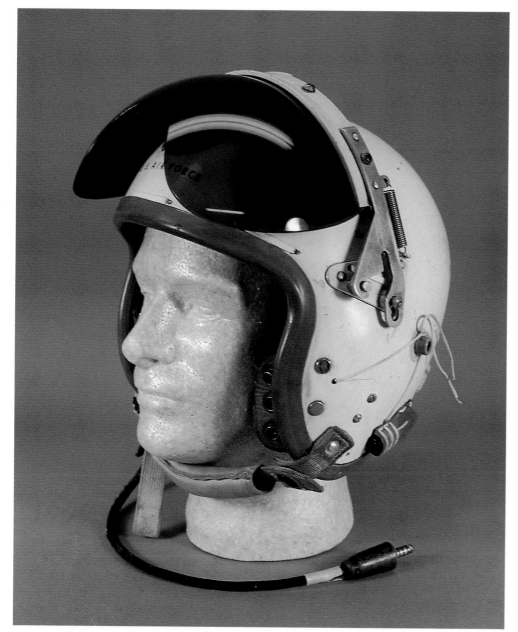
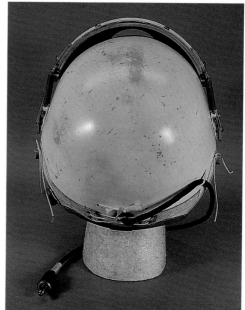
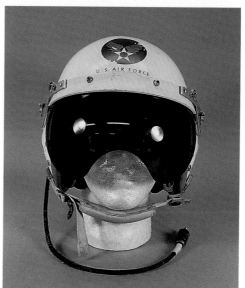

The U.S.A.F. P-4 helmet was a continuation of the P-3 with improved avionics. It was standardized in 1957 and became obsolete in 1961. This P-4 helmet was made by the Selby Shoe Company.

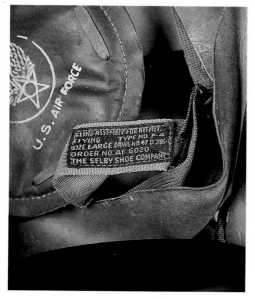
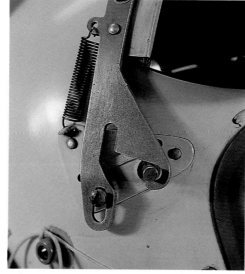

Visor mechanism - Up position.

Visor mechanism - Down and locked.

P-4A

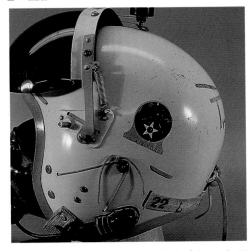

The large wing and star insignia decal of the early 'P' series helmets was reduced in size and moved to the left side of the helmet.

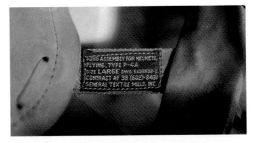

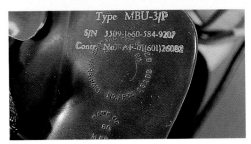

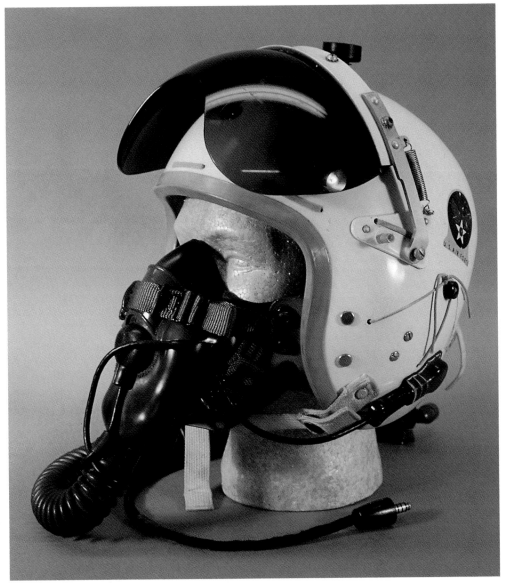

U.S.A.F. P-4A helmet with external visor and MBU-3/P oxygen mask. This P-4A, standardized in 1956, was used in fighter and bomber aircraft through the 1960s and declared obsolete in 1964. These helmets were issued with external visors that used a center locking track. The MBU-3/P oxygen mask is a U.S.A.F. redesignation of the MS22001.

The CRU-8/P oxygen connector replaced the MC-3A. The small goose neck fitting is used for attachment to the emergency bailout oxygen system.

The U.S.A.F. insignia is used on the inner suspension system on 'P' helmets.

CHAPTER 1: U.S.A.F./U.S. ARMY FLIGHT HELMETS & HIGH ALTITUDE HELMETS

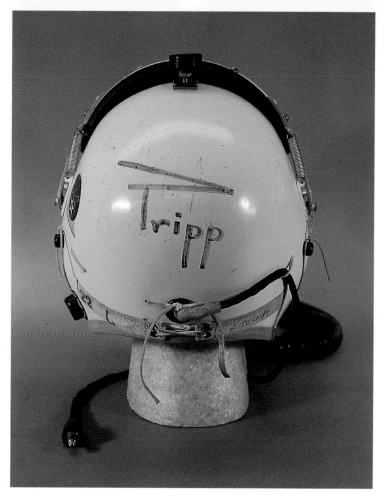

P-4B

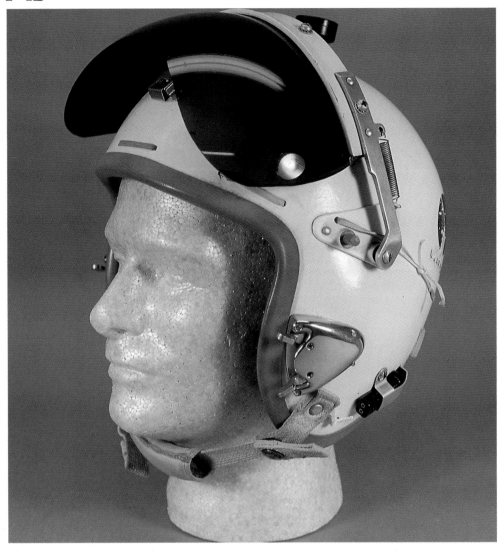
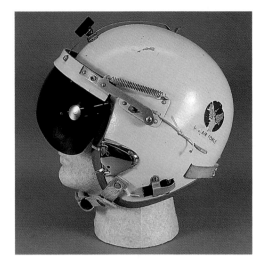
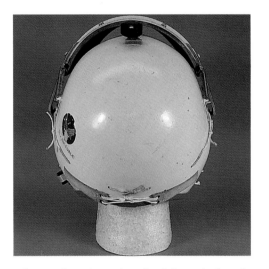

The P-4B helmet is similar to the P-4A with minor modifications. The U.S.A.F. P-4B has chromed 'Hardman' type external oxygen bayonet receivers fitted. Some also have the cast aluminium receivers. This P-4B helmet was made by Consolidated Controls Corporation.

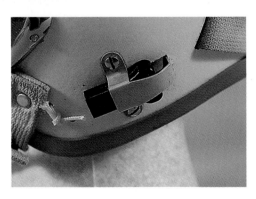
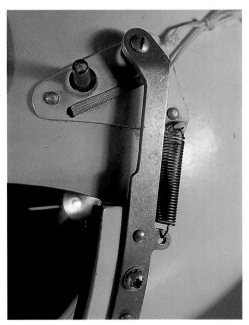
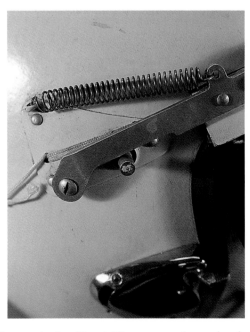

ABOVE LEFT: Visor mechanism 'Up.' ABOVE RIGHT: Visor mechanism 'Down' (The center track served as the locking mechanism). LEFT: Oxygen microphone attachment bracket (Earlier 'P' helmets had this fitting covered in either leather or rubber).

CHAPTER 1: U.S.A.F./U.S. ARMY FLIGHT HELMETS & HIGH ALTITUDE HELMETS

MB-3

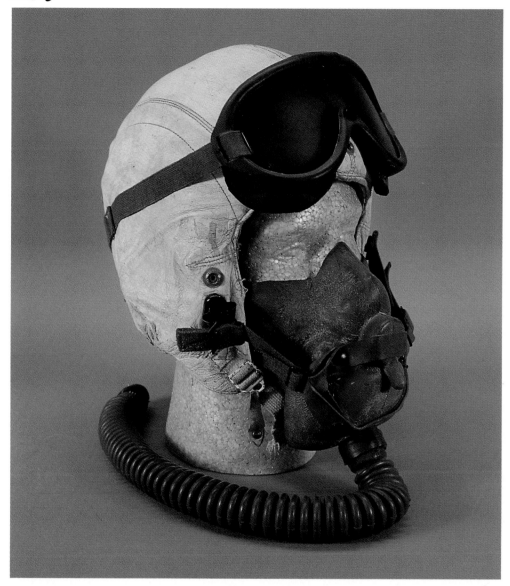

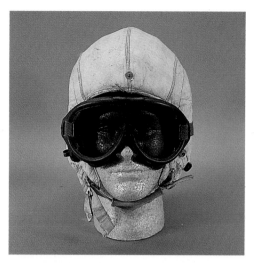

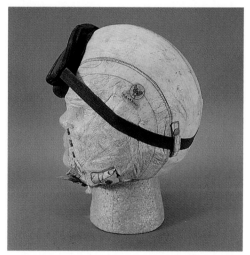

U.S.A.F. MB-3 helmet with MBU-4/P oxygen mask and M-1944 goggles. This helmet, made by Aviators Equipment Corporation, was used by the Strategic Air Command in the late 1950s, early 1960s for crew positions not requiring use of a rigid helmet. The MB-3 is constructed of a laminated inner shell covered in white leather. A flexible bill could be attached to the helmet but was rarely used.

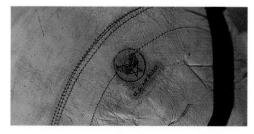

A small U.S.A.F. decal is placed on the left side of the helmet.

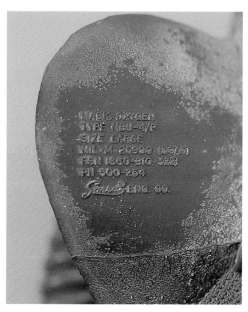

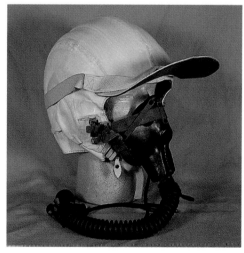

LEFT: MBU-4/P oxygen mask nomenclature (The MS22001 mask was also used).

19

MB-4

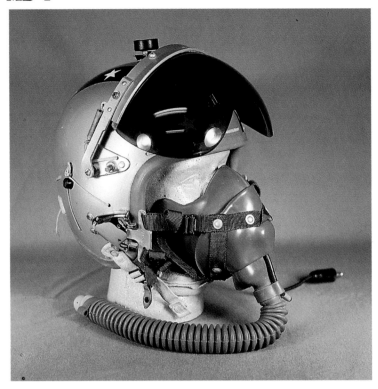
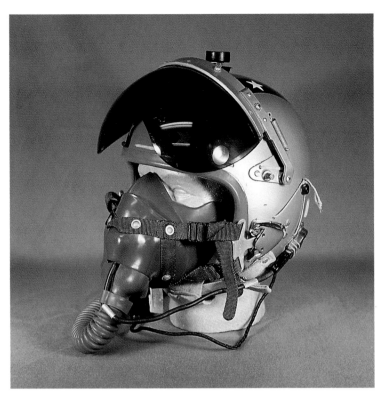

U.S.A.F. MB-4 helmet with external visor and MBU-3/P oxygen mask. This helmet, made by the Selby Shoe Company, was used by the Strategic Air Command in the mid 1950s and early 1960s in such aircraft as the B-47 Stratojet. The MB-4 helmet is similar to the P-4A and B helmets and uses 'Hardman' receivers and double slot 'Hardman' bayonets.

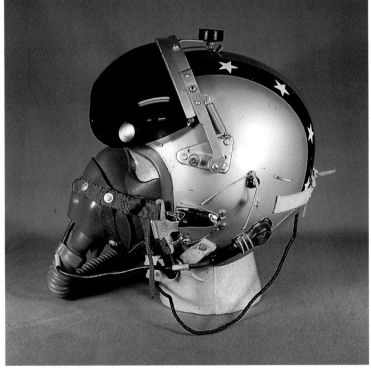

CHAPTER 1: U.S.A.F./U.S. ARMY FLIGHT HELMETS & HIGH ALTITUDE HELMETS

HGU-2/P

U.S.A.F. HGU-2/P helmet with enclosed single visor housing, tinted visor and MBU-5/P oxygen mask. Standardized in 1957, the HGU-2/P was the first U.S.A.F. helmet issued with a visor housing and was superseded by the HGU-2A/P helmet in 1960. It was used through the 1960s in fighter and bomber type aircraft. The shell construction of the helmet is a fiberglass cloth reinforced with epoxy resin. The visor locking knob secured the visor 'up' or 'down' by turning until tight.

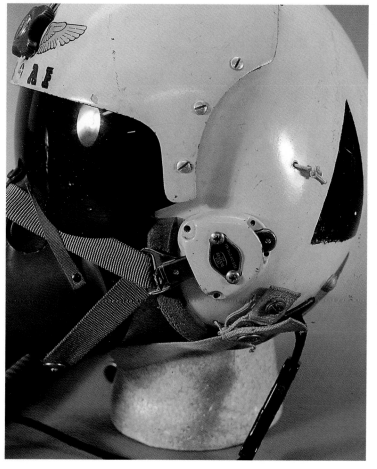

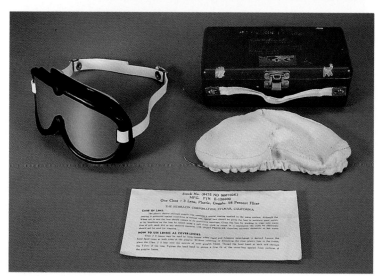

ABOVE: Nuclear flash goggles (M-1944) were used with the P-4, HGU-2/P and 2A/P.

RIGHT: Single slot oxygen bayonet and cast aluminium receiver.

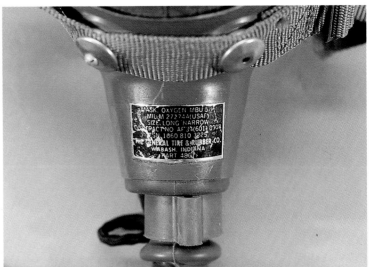

LEFT: CRU-8/P oxygen connector. RIGHT: MBU-5/P oxygen mask was made by various manufacturers including General Tire & Rubber, Sierra and Gentex. This MBU-5/P mask uses an M100/A1C microphone.

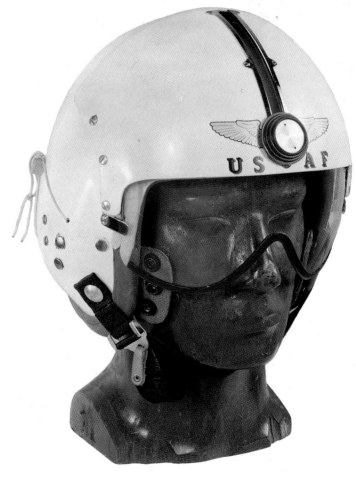

Earlier HGU-2/P helmets used a round visor knob with a push button locking device and leather tabs and snaps for securing the oxygen mask.

CHAPTER 1: U.S.A.F./U.S. ARMY FLIGHT HELMETS & HIGH ALTITUDE HELMETS

HGU-2A/P

U.S.A.F. HGU-2A/P helmet with an MBU-5/P oxygen mask.

ABOVE: Double slot 'T' oxygen bayonet and cast aluminum receiver. BELOW: The CRU-60/P oxygen connector replaced the CRU-8/P.

ABOVE: A gold tinted visor was sometimes used for protection from the high intensity illumination resulting from nuclear detonations as well as other bright glare sources. A neutral grey and clear visor were also available.

FAR LEFT: The visor housing leading edge has been modified for added visibility. The shell is constructed of a fiberglass cloth reinforced with epoxy resin.

LEFT: This Gentex HGU-2A/P with MBU-5/P oxygen mask and offset bayonets was used in an F-15E Eagle in 1981. The leading edge of the visor housing has been slightly cut away to enhance visibility. Note how chamois cloth has been added to the visor locking knob and the top of the helmet to protect the canopy from scratches during 'G' maneuvers. The Sierra Engineering Company also manufactured the HGU-2A/P.

23

APH-5A

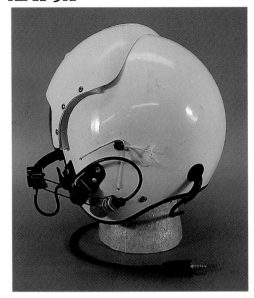

The shell is constructed of a fiberglass cloth reinforced with epoxy resin.

Gentex APH-5A helmet in white.

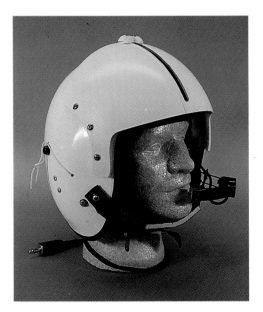

U.S. Army APH-5A Vietnam era helicopter and reconnaissance aircraft (OV-1) helmet in olive drab with boom mounted M33A/A1C microphone.

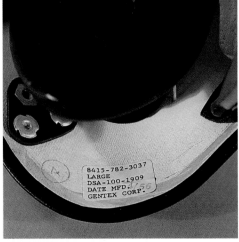

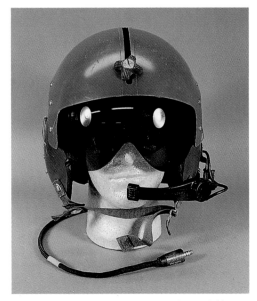

Neutral grey visor (A clear visor was also available).

AFH-1

U.S. Army AFH-1 Vietnam era helicopter and reconnaissance aircraft (OV-1) helmet in olive drab with clear visor, boom mounted M87/A1C microphone and HGU-4/P sunglasses. The shell is constructed of a phenol formaldehyde and polyvinyl butyryl resin combination bonding a nylon ballistic cloth.

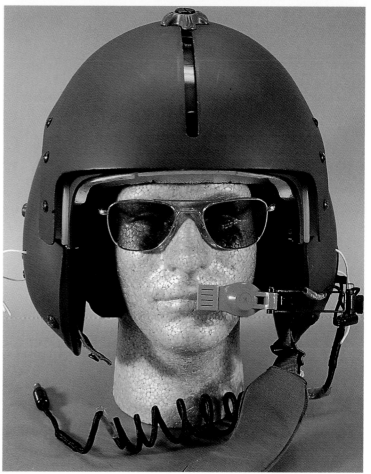

RIGHT: These helmets were made by Sierra Engineering Company and Gentex Corporation.

SPH-4

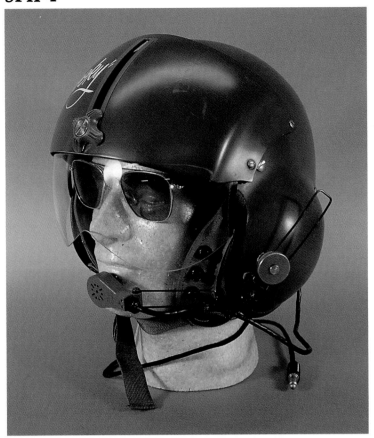
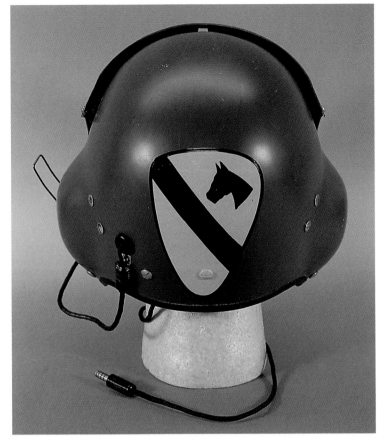

U.S. Army SPH-4 Vietnam era helicopter helmet in olive drab with clear visor, boom mounted M138/U microphone, and HGU 4/P sunglasses. The Gentex SPH-4 helmet uses a six-point attachment adjustable cotton/nylon suspension system. Snaps are provided on each side of the retention assembly to fasten the chin strap or optional oxygen mask. The shell is constructed of fiberglass cloth reinforced with epoxy resin. A neutral grey visor was available. Note U.S. Army 1st Cavalry (Airmobile) insignia above right.

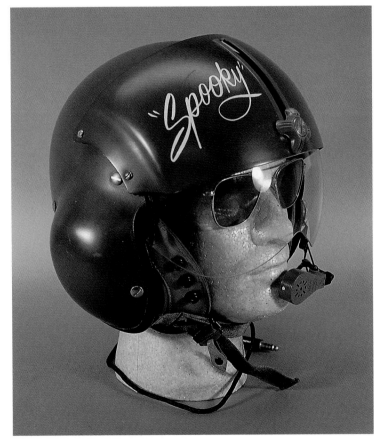
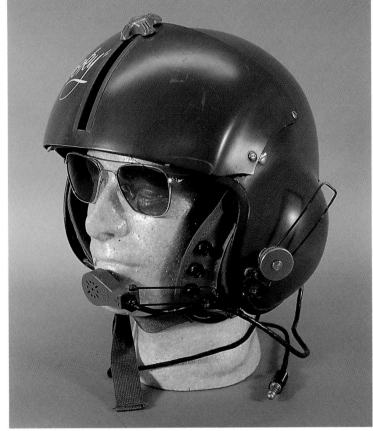

CHAPTER 1: U.S.A.F./U.S. ARMY FLIGHT HELMETS & HIGH ALTITUDE HELMETS

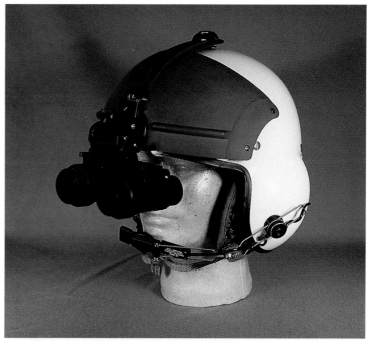

A 1990s U.S.A.F. SPH-4 helmet fitted with an ANVIS model M-927 night vision goggle (NVG) and boom mounted M87/A1C microphone (This helmet was used in a UH-1N Huey helicopter).

LEFT: A 1990s U.S. Army Cobra gunship SPH-4 helmet fitted with a monocle type sighting system. RIGHT: U.S. Army sighting system mounting assembly showing label and aircraft to helmet electrical connection.

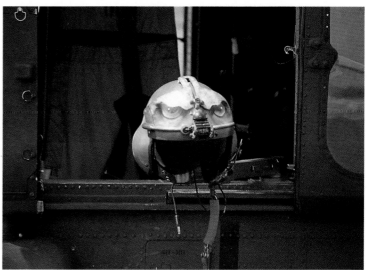

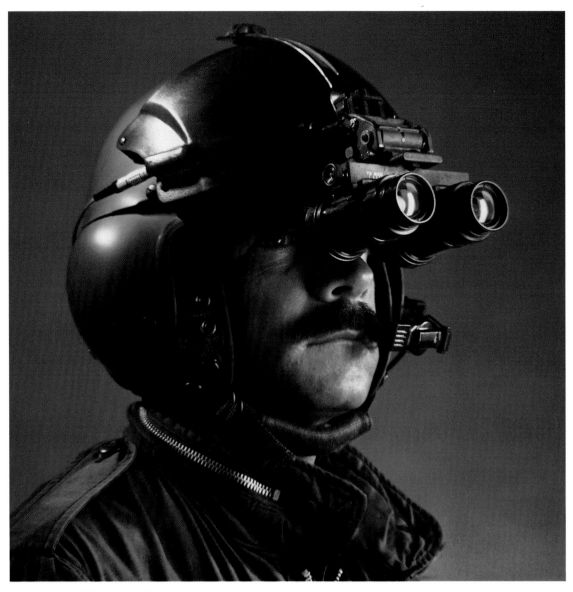

NVG mounting bracket. Note how the NVG electrical lines are routed through build-in channels on either side of the visor housing. The visor locking knob extension on the visor cover is required because of the position of NVG mount.

CHAPTER 1: U.S.A.F./U.S. ARMY FLIGHT HELMETS & HIGH ALTITUDE HELMETS

H79/A1C HEADSET

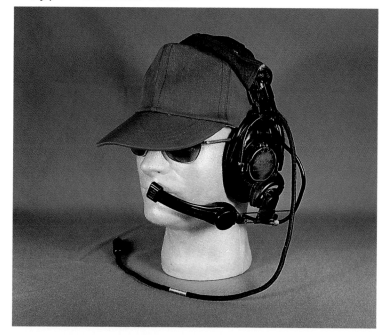
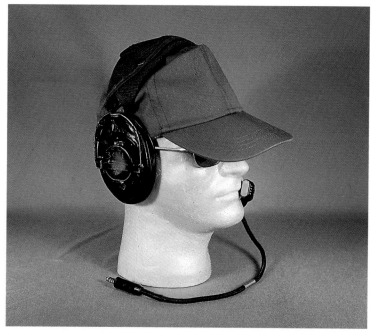

U.S.A.F. H79/A1C headset with boom mounted M33/A1C microphone, HGU-4/P sunglasses and standard utility cap. This was used in such aircraft as the O-1G Birddog in 1967 and was replaced by the H-157/A1C headset.

H-157/A1C HEADSET

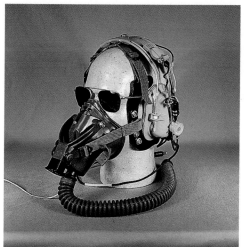

ABOVE: U.S.A.F. H-157/A1C headset with boom mounted M-87/A1C microphone, quick don type P/N 450-360 MBU-5/P oxygen mask and HGU-4/P sunglasses. This was a standard arrangement for cargo and tanker type aircraft. The boom microphone (shown in 'up' stowed position) could be attached on either side of the headset. A cloth oxygen mask cover is shown.

RIGHT: Author Mike Breuninger in Southeast Asia in 1971 using the H-157/A1C headset in a C-130E transport.

29

HGU-7/P

The MBU-5/P oxygen mask attaches by snaps.

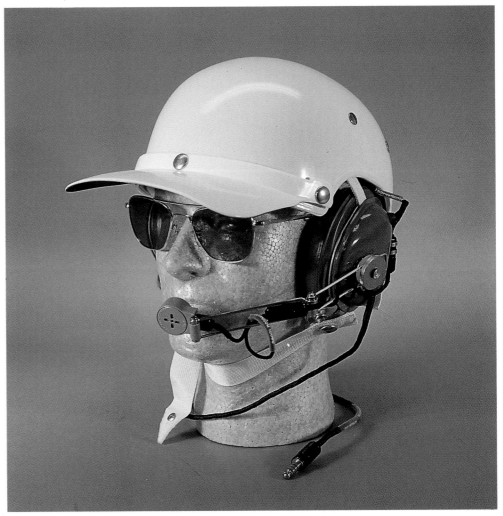

U.S.A.F. HGU-7/P helmet with boom mounted 11110 microphone, 256/A/C headset, removable sun bill and HGU-4/P sunglasses. The HGU-7/P, standardized in 1959 and made by Land Manufacturing, Inc., and Gentex Corporation, was used by crew members of F.A.C., cargo, trainer, and tanker type aircraft and replaced the MB-3 helmet. The helmet is constructed of an injection molded polycarbonate material. M-1944 goggles could also be used.

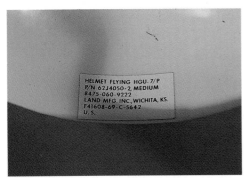

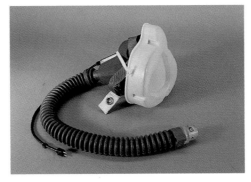

A plastic cover is used to protect the mask from dirt when not in use.

RIGHT: Oxygen mask microphone M101/A1C.

CHAPTER 1: U.S.A.F./U.S. ARMY FLIGHT HELMETS & HIGH ALTITUDE HELMETS

'TOPTEX" 3BM

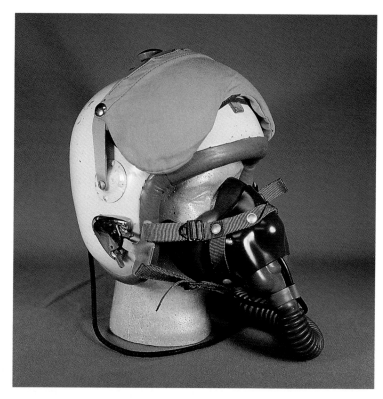

U.S.A.F. 'TOPTEX' 3BM helmet with MS22001 oxygen mask and visor cover. This helmet produced by Protection, Inc. Division of Mine Safety Appliance Company was typically used in aircraft such as the TF-102 Delta Dagger and by many test pilots in the 1960s and 1970s. The 3BM uses 'Hardman' receivers and a center visor locking track similar to the P-4B helmet. The oxygen mask held the helmet on because a chin strap was not used.

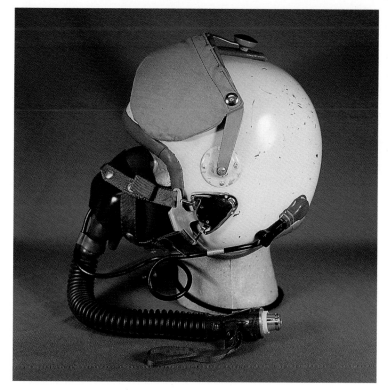

'TOPTEX" 3BM

A 3BM helmet with an MBU-3/P was used by the U.S.A.F. Thunderbirds in 1959 in the F-100D Super Sabre.

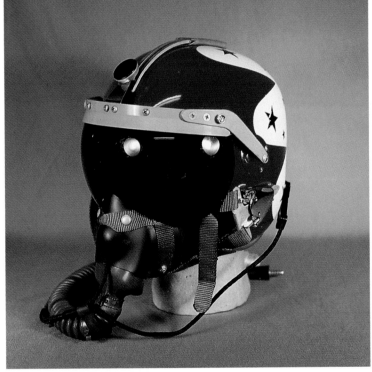

CHAPTER 1: U.S.A.F./U.S. ARMY FLIGHT HELMETS & HIGH ALTITUDE HELMETS

HGU-16A/P

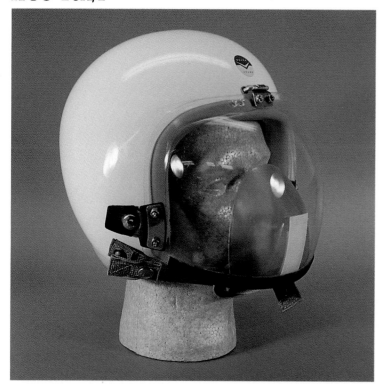
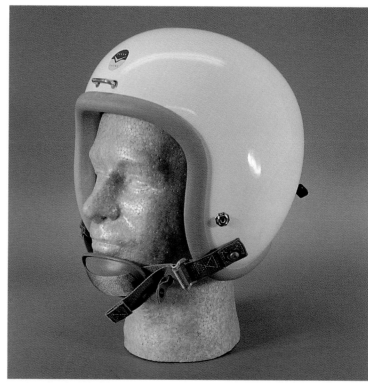

U.S.A.F. HGU-16A/P para-rescue helmet with protective face shield.

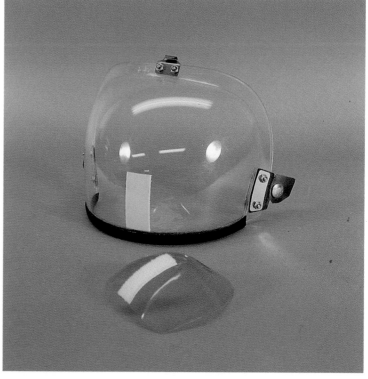

Face shield with anti-fog breath deflector removed.

HGU-26/P

U.S.A.F. HGU-26/P dual visor helmet with MBU-5/P oxygen mask and neutral grey visor. The HGU-26/P helmet, manufactured by Sierra Engineering Company and Gentex Corporation, was introduced in the early 1970s. Its dual visor allowed use in various fighter and bomber aircraft. The shell of the HGU-26/P was also used in the HGU-2A/P helmet and is constructed of fiberglass cloth reinforced with epoxy resin. The MBU-5/P oxygen mask was introduced in the mid-1960s and replaced the Air Force MS22001 mask.

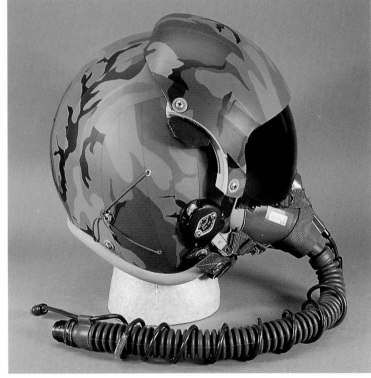

Clear visor.

CHAPTER 1: U.S.A.F./U.S. ARMY FLIGHT HELMETS & HIGH ALTITUDE HELMETS

A visor locking knob is located on each side, one for each of the two visors. The visor housing is known as a PRU-36/P assembly.

Offset oxygen bayonet and receiver were first used in the early to mid 1970s.

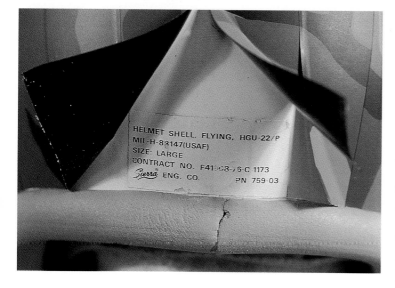

RIGHT: A grey cloth skull cap is worn next to the head to absorb perspiration and enhance comfort. Skull caps were first worn in the early 1970s and were available in a variety of colors.

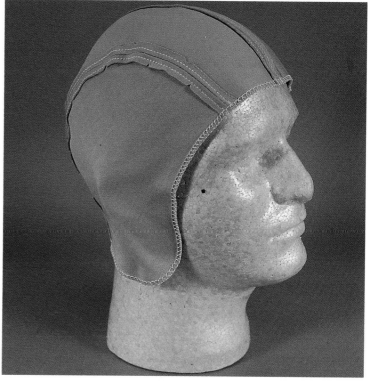

SIERRA HALO

A 1974 Sierra Engineering Co. HALO (High Altitude Low Opening) helmet with two-piece clear face shield and Sierra oxygen mask. LEFT: Upper face shield attached.

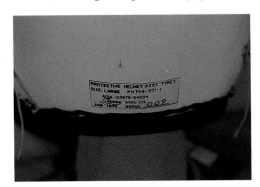

The oxygen mask is attached to leather tabs.

EDP-2/P

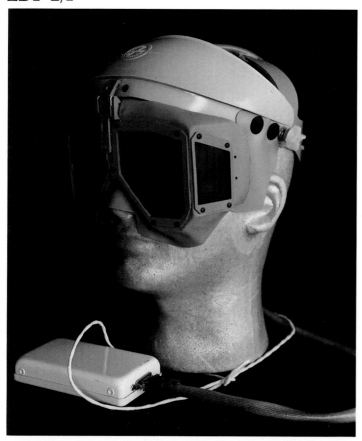
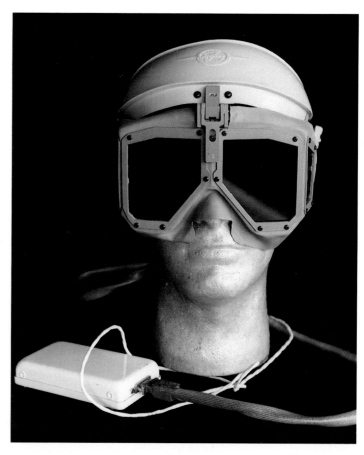

U.S.A.F. EDP-2/P nuclear flash protection goggles. The EDP-2/P goggles were worn by SAC KC-135 crews in the late 1970s and are the same unit used in the HGU-51/P Nuclear, Biological, Chemical (NBC) helmet. When charged with 26 VDC, the goggles become clear and only moderately tinted. Sensors located within the goggles react to bright light, such as the blinding flash created by a nuclear detonation, and in a nanosecond, the lenses are darkened to that of arc-welding goggles. The use of the goggles was classified secret for many years. The electronics are contained in the white control box.

Protective storage box.

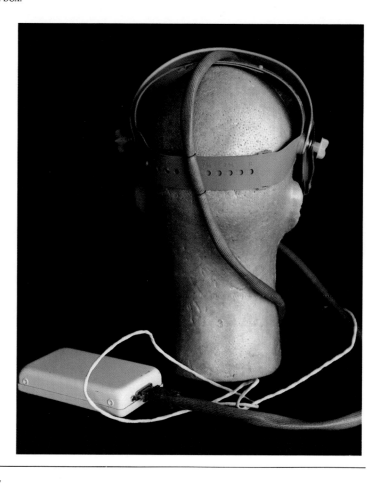

MBU-10/P

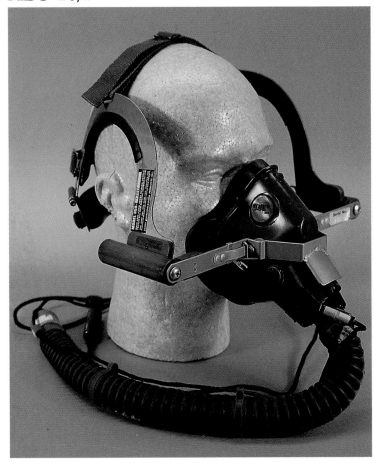
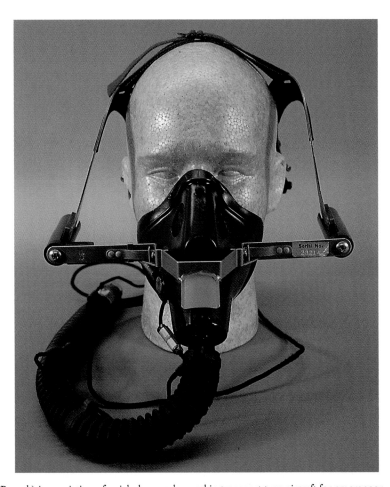

This U.S.A.F. type MBU-10/P quick-don oxygen mask assembly (utilizing an MBU-5/P mask) is a variation of quick-don masks used in transport type aircraft for emergency oxygen requirements. The standard type H-157/A1C headset is used. It has been in use since the late 1970s.

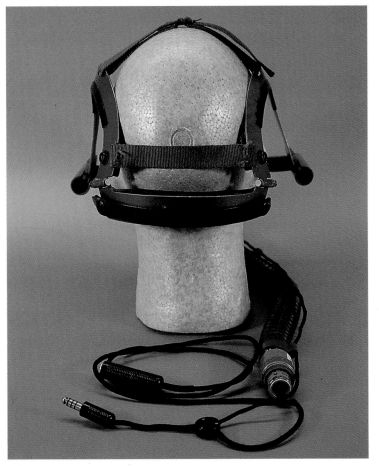

CHAPTER 1: U.S.A.F./U.S. ARMY FLIGHT HELMETS & HIGH ALTITUDE HELMETS

HGU-39/P

U.S.A.F. HGU-39/P chemical, biological, oxygen helmet and mask used in the 1980s. The basis of the HGU-39/P is an SPH-4 helmet without the visor cover. Cast aluminum oxygen receivers and Air Force communication components complete the basic helmet. RIGHT: The sun visor attaches to the mask with a Velcro strap at each side.

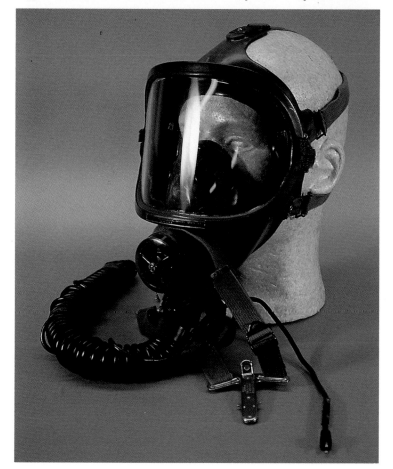

The chemical hood is placed over the helmet and mask.

LEFT: The mask is donned first and uses double slot 'T' bayonets to secure it to the helmet.

  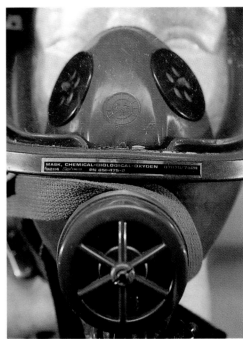

CRU-80/P filter/oxygen assembly.

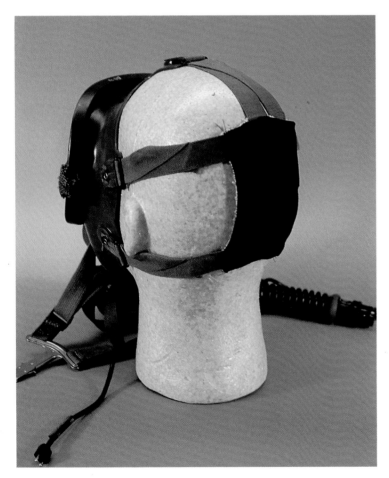 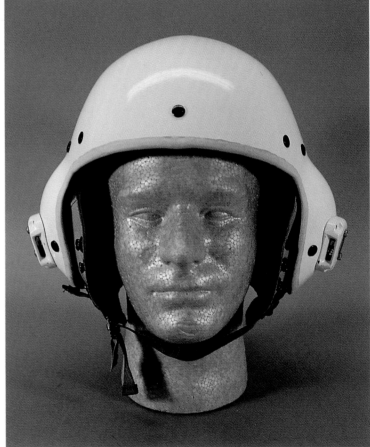

CHAPTER 1: U.S.A.F./U.S. ARMY FLIGHT HELMETS & HIGH ALTITUDE HELMETS

HGU-51/P

The face/neck skirt is attached directly to the clear visor.

The left oxygen mask bayonet incorporates the internal ventilation duct.

U.S.A.F. HGU-51/P Integrated Chemical Defense Helmet System with EDP-2P nuclear flash protection goggles, clear visor, and face/neck skirt covering the oxygen mask. The HGU-51/P NBC helmet system, made by the Gentex Corporation, was procured in limited numbers primarily for F-111 pilots in Europe in the late 1970s.

The left oxygen hose provides ventilation for the helmet and visor and is controlled by a compound switch on the left side. The right hose provides oxygen to the mask through internal ducting. The mask has the same drinking tube system as the M-17A1 respirator.

HGU-55/P

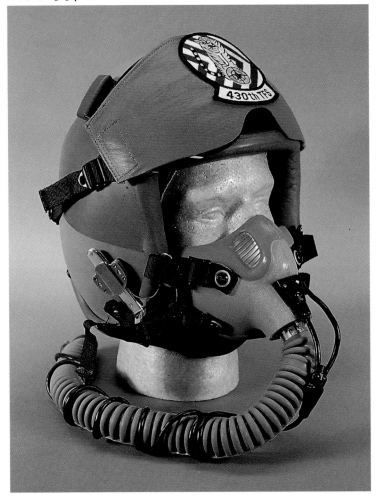

U.S.A.F. HGU-55/P helmet with MBU-12/P oxygen mask and leather visor cover. The Gentex Corporation HGU-55/P was introduced in 1983 and is still one of the most widely used helmets in service with the U.S.A.F., U.S.N., U.S.M.C. and foreign countries. It is used in fighter, bomber, and some specialized transports such as the C-130 gunship. The helmet is designated as the HGU-36/P when used in the C-130 gun ship and has a ballistic shell. The HGU-55/P shell is made of fiberglass cloth reinforced with epoxy resin. A dual visor system is available without a housing as well as the PRU-36/P side actuated dual visor assembly used on the HGU-26/P helmet. The MBU-12/P oxygen mask was introduced in 1981 to replace the U.S.A.F. MBU-5/P and U.S.N. A13-A (MS22001) masks and uses an M-101/A1C microphone. The first MBU-12/P oxygen masks were green and the grey colored mask came into service in the mid 1980s. Note the insignia of 430th Tactical Fighter Squadron.

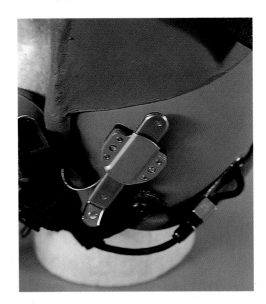

Lightweight bayonet receiver.

The leather visor cover has been removed. The neutral grey visor is secured by an elastic strap and snap on each side of the helmet.

CHAPTER 1: U.S.A.F./U.S. ARMY FLIGHT HELMETS & HIGH ALTITUDE HELMETS

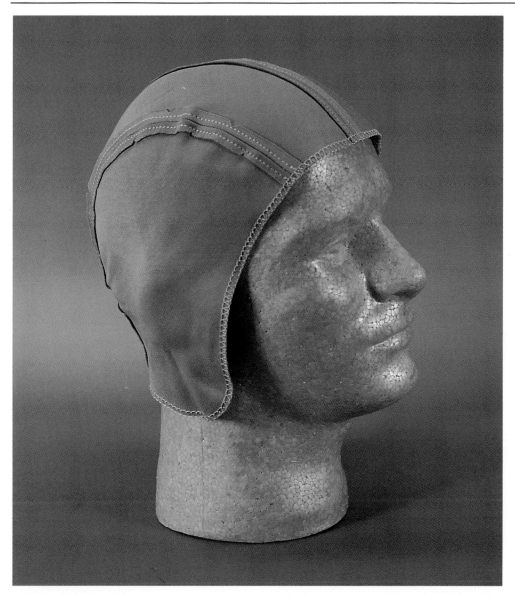

Grey cotton skull cap.

Skull cap label.

An amber colored visor for high light and vermillion colored visor for standard sun protection are shown. A dark yellow (low light), clear and various laser visors (varying in color from light green to dark brown) are also available.

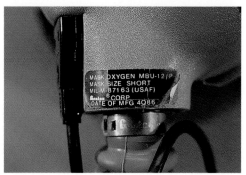

CRU-60/P oxygen connector.

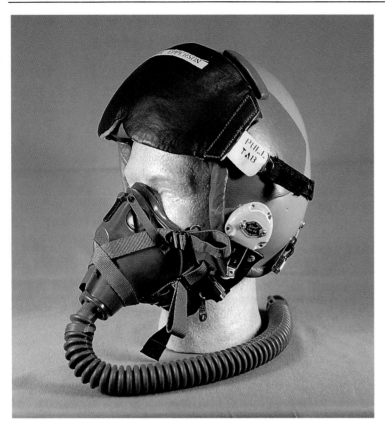 

Early HGU-55/P with cast aluminum bayonet receivers and MBU-5/P mask. S.A.C. HGU-55/P dual visor helmet with mounting plate for the EEU-2A/P nuclear flash protection goggles used in the late 1970s and 1980s in B-52, F-111, KC-135, and B-1B aircraft. It is shown with the MBU-5/P oxygen mask. The operation is the same as the EDP-2/P goggles.

CHAPTER 1: U.S.A.F./U.S. ARMY FLIGHT HELMETS & HIGH ALTITUDE HELMETS

EEU-2A/P flash goggles mounted on the HGU-55/P.

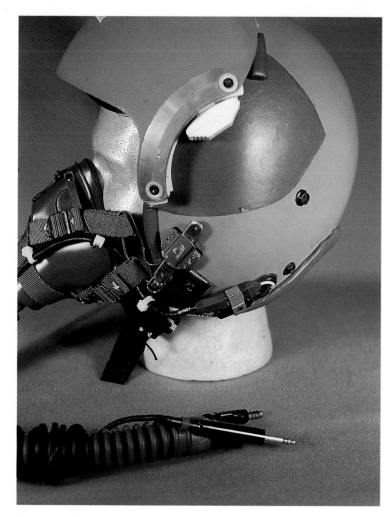

Oxygen mask microphone and EEU-2A/P goggle hook up.

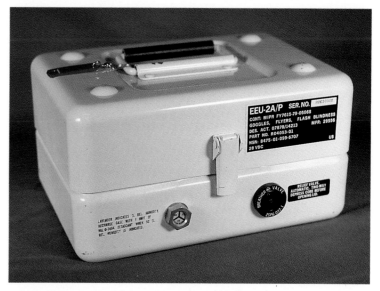

Storage/transport container for the EEU-2A/P goggles.

45

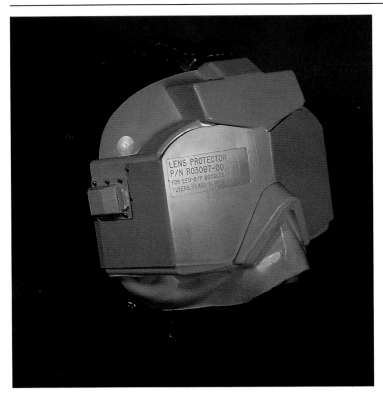

Lens protective cover used in storage and transport.

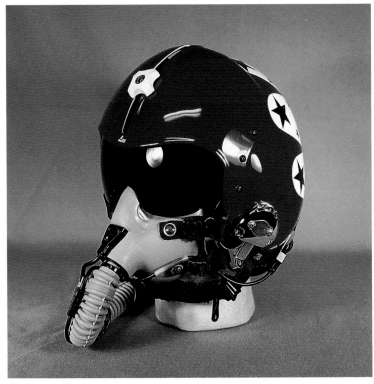

ABOVE AND TWO BELOW: This U.S.A.F. F-16 Thunderbird helmet uses an enclosed visor housing and MBU-12/P mask.

CHAPTER 1: U.S.A.F./U.S. ARMY FLIGHT HELMETS & HIGH ALTITUDE HELMETS

SPH-4B & 4AF

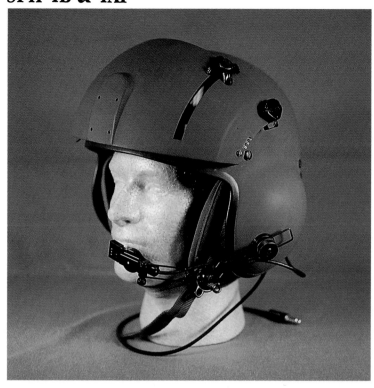

U.S. Army 1990s issue SPH-4B dual visor helicopter helmet in olive drab with boom mounted M-87A/A1C microphone. The visor housing provides a mounting platform for the AN/VIS-6 night vision goggles.

1990s issue U.S.A.F. SPH-4AF helicopter helmet with clear lens. Other special visor tints are available for laser protection, high contrast, etc. Note the black colored retention system used on the U.S.A.F. helmet versus the olive drab colored retention system used on the U.S. Army helmet. The Gentex Corporation produced both the U.S.A.F. SPH-4AF and the U.S. Army SPH-4B.

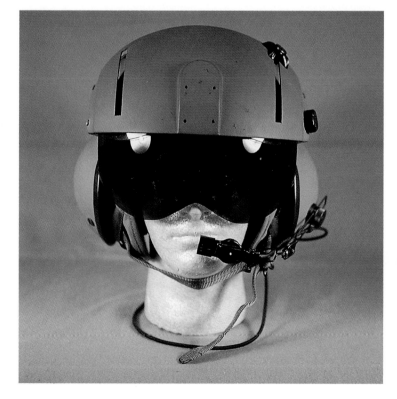

U.S. Army SPH-4B helmet with neutral grey visor.

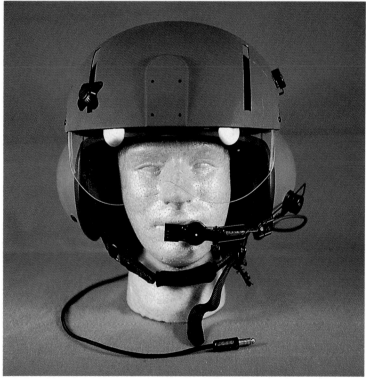

SPH-4AF helmet.

IHADSS/AB-05 (Integrated Helmet and Display Sighting System)

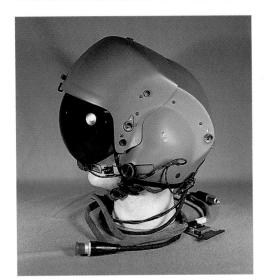

U.S. Army 1990s IHADSS/AB-05 (Integrated Helmet And Display Sighting System) AH-64 Apache helicopter helmet with sight and boom mounted M170/A1C microphone.

IHADSS/AB-05 helmet with neutral grey visor. Note the visor cut out to accommodate the sighting system.

CHAPTER 1: U.S.A.F./U.S. ARMY FLIGHT HELMETS & HIGH ALTITUDE HELMETS

T-1 HALO

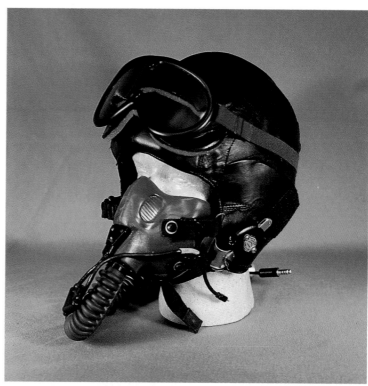

U.S. Army HALO (High Altitude Low Opening) leather parachutists helmet with M-1944 goggles and MBU-12/P oxygen mask. The T-1 is used by special operations groups for high altitude jumping from C-141 and C-130 aircraft. This example is from 1989.

Offset bayonets are inserted into cast aluminum receivers. An M-101/A1C microphone is used in this oxygen mask.

BELL HALO

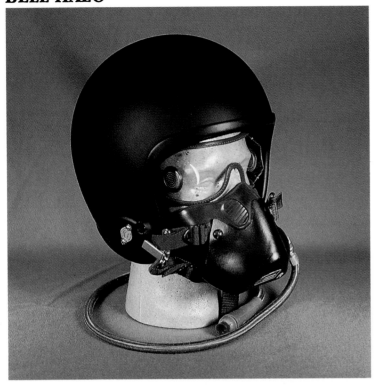

1991 Bell HALO parachutists helmet with free fall goggles and R-267/12-P oxygen mask. Double slot bayonets are inserted into cast aluminum receivers.

CHAPTER 1: U.S.A.F./U.S. ARMY FLIGHT HELMETS & HIGH ALTITUDE HELMETS

GENTEX HALO

Gentex HALO Air Force 'Parachutists' helmet with boom mounted M97/A1C microphone, external detachable visor and clear goggles. The Gentex HALO helmet is also used by U.S. Army special operations groups and foreign countries. The boom microphone can be positioned on either side of the helmet. The helmet uses lightweight oxygen mask receivers for either the MBU-5/P, MBU-12/P on the specially designed Special Operations Command SOCOM(HA-LP) oxygen masks. The helmet is shown painted in lusterless grey but is also available in flat black.

HGU-53/P

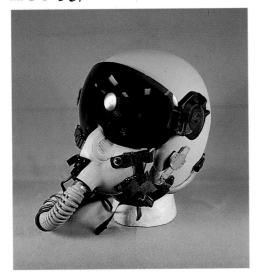

U.S.A.F. HGU-53/P helmet with MBU-12/P oxygen mask and visor locking system. The Gentex Corporation HGU-53/P was the first 600 knot U.S. helmet and was developed to replace the HGU-55/P. The 53/P, however, does not integrate with all the equipment used with the 55/P helmet such as the flash blindness goggles. The helmet shell is made of a light weight Kevlar material.

Continuous locking rotary visor system.

LEFT: The profile of the HGU-53/P differs from previous helmets with a reduction in external dimensions for improved canopy clearance. RIGHT: Leather visor cover installed.

HGU-55/P (CE)

U.S.A.F. HGU-55/P (CE) 'Combat Edge' helmet with MBU-20/P oxygen mask and clear visor. The Gentex Corporation HGU-55/P (CE) 'Combat Edge' helmet is part of a tactical total pressure breathing system being used today in F-15 and F-16 fighter aircraft. The system improves high performance fighter pilot tolerance to high G maneuvers up to +9 G's. The helmet is used in conjunction with a low profile MBU-20/P side hose oxygen mask, CSU-17/P counter pressure vest, CRU-94/P oxygen terminal block and CSU-13B/P anti-G pants. The helmet utilizes a urethane-coated nylon bladder installed between the thermo plastic liner (TPL) and the energy absorbing liner in the rear of the helmet. The bladder is connected to the MBU-20/P oxygen mask and interfaces with the pressure breathing system. During high G maneuvers the bladder inflates to provide automatic mask tensioning to lock the mask into position. Other visor tints similar to those used with the HGU-55/P are available.

JET AGE FLIGHT HELMETS

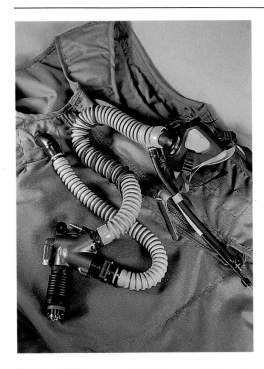

The MBU-20/P oxygen mask is shown attached to the CRU-94/P terminal block and CSU-17/P counter pressure vest. The MBU-20/P mask uses an M-169/A1C microphone.

CRU-94/P.

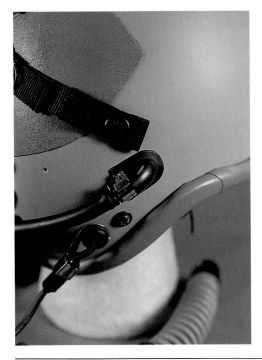

This pilot is wearing the 'Combat Edge' HGU-55/P (CE) helmet, MBU-20/P oxygen mask and CRU-94/P terminal block. The CSU-17/P counter pressure vest is being worn under the SRU-21/P survival vest.

LEFT: MBU-20/P oxygen mask microphone connection (lower line) and helmet air bladder quick disconnect line.

HGU-56/P

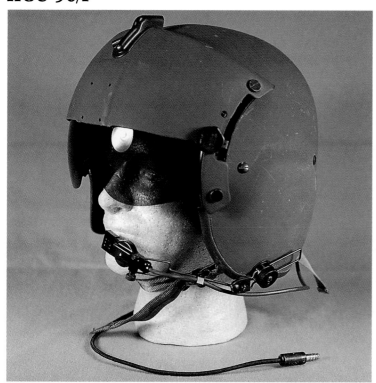
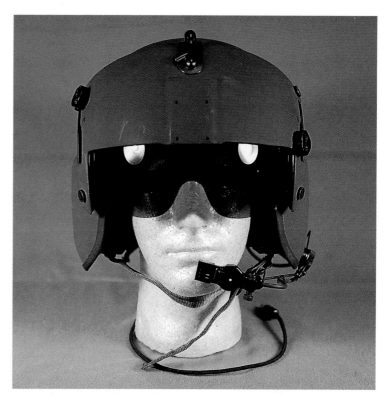

Current issue Gentex Corporation U.S. Army HGU-56/P dual visor helicopter helmet with neutral gray visor, boom mounted M-87A/A1C microphone, and visor housing platform to accept night vision goggles (NVG). The HGU-56P replaces the SPH-4 and 4B helmet and provides the U.S. Army with one helmet for all aircraft through a modular visor housing platform system. Visor housings are available for equipment applications in Apache, Cobra, and Blackhawk helicopters. The HGU-56/P helmet is compatible with the U.S. Army M-43 chemical warfare protection mask and hood and can also be adapted to allow use of the standard U.S.A.F. oxygen mask for high altitude operation. The helmet shell is fabricated from an advanced graphite composite for lighter weight and impact protection.

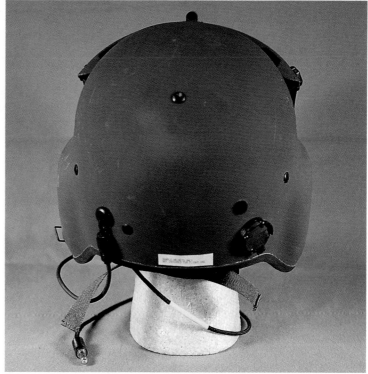

An HGU-56/P helmet fitted with an Apache helicopter visor and sighting system.

CHAPTER 1: U.S.A.F./U.S. ARMY FLIGHT HELMETS & HIGH ALTITUDE HELMETS

HIGH ALTITUDE HELMETS
(PARTIAL-PRESSURE AND FULL-PRESSURE)

S-1

This handmade S-1 helmet and suit was the first high-altitude partial-pressure outfit made by the University of Southern California's Dr. James Paget Henry in 1945.

T-1

U.S.A.F./N.A.C.A. high-altitude partial-pressure helmet worn by Chuck Yeager in the X-1 on his record-breaking Mach 1 flight in 1947. This handmade helmet was sealed by a thin rubber hood worn inside and adjusted tightly around the neck like a collar. A nylon outer hood supported the rubber inner hood and was also adjusted to fit snugly with laces and a zipper. The helmet had an electric defogging/defrosting face shield utilizing tiny wires to heat the plexiglass. It was designed to seal around the face but the face shield which was sealed to the helmet created many comfort problems. To be effective, the helmet had to be worn tightly, making the wearer very uncomfortable in less than one hour. The helmet was for use with the T-1 partial-pressure suit. BELOW: This dog helmet/oxygen mask combination is being worn by a Saint Bernard for flight and parachute tests. It was designed by Alice King Chatham, the designer behind the T-1 helmet worn by Chuck Yeager.

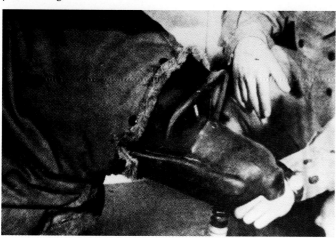

Fiberglass protective shell.

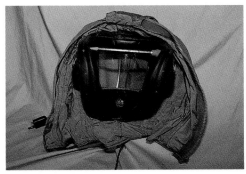

Production version T-1 high altitude partial-pressure helmet. The T-1A helmet, first produced in 1947 by Mine Safety Appliances Co., was used by the U.S.A.F. and N.A.C.A. in X-1 flights. Standard avionics and a snap-on fiberglass protective shell are used with the helmet. It remained very unpopular due to the lack of comfort and a fixed face shield that nearly caused at least one crash when the pilot could not effectively clear his ears, causing severe pain. In the photos, the inner rubber hood is missing due to deterioration.

Interior hood showing integrated avionics and fixed face shield.

The face shield is permanently attached to the hood.

CHAPTER 1: U.S.A.F./U.S. ARMY FLIGHT HELMETS & HIGH ALTITUDE HELMETS

K-1 'SPLITSHELL'

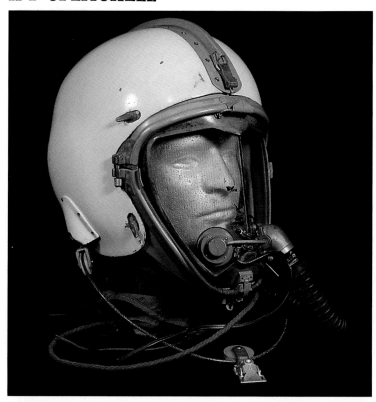

K-1 'Splitshell' high-altitude partial-pressure helmet. The K-1 'Splitshell', produced in the late 1940s by International Latex Corp. (ILC), was an evolution of the original 'Chatham' helmet design with many improvements. To improve comfort, the aluminium frame was padded and covered with soft leather, as were the earphones. Soft padded leather also replaced the critical rubber face seal. The metal outer frame added structural integrity for comfort, crash protection, and durability. It also provided a solid mount for the microphone and earphones. The white fiberglass outer protective shell is constructed in two halves, hence the nickname 'Splitshell.' They are held in place by angled mounting studs and a center flange which clips at the front and holds them rigid. The face plate is removable and is made from two flat pieces of tempered glass held in place by a hard rubber frame. It is electrically heated and has the pressure-demand oxygen system mounted. Due to the helmet's weight, it was the first to use a negative-G hold-down arrangement which connected the helmet to the T-1 partial-pressure suit with a cable and pulley system to reduce the force on the pilot's neck during high negative G's. The 'Splitshell' was used in test flights of many types of aircraft of the early jet age.

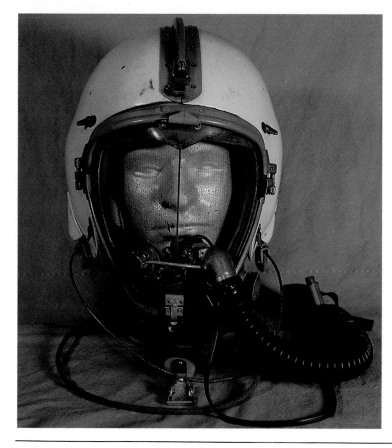

Hood/neck skirt with rigid metal frame.

Interior of hood showing leather covered earphones and padding.

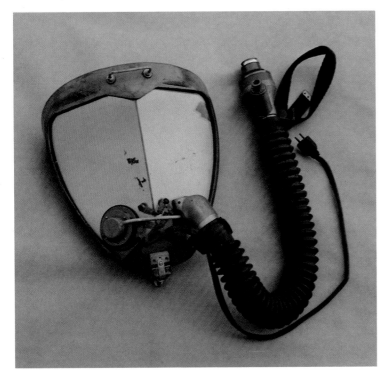

K-1 face plate.

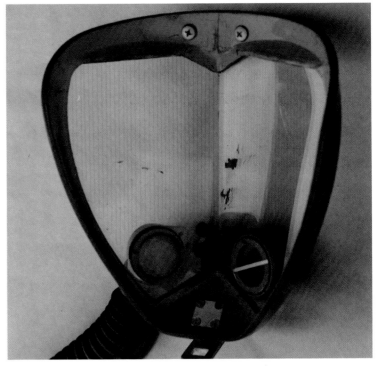

K-1 face plate (interior).

CHAPTER 1: U.S.A.F./U.S. ARMY FLIGHT HELMETS & HIGH ALTITUDE HELMETS

LEFT: The emergency bailout bottle arrangement pressurized the suit as well as provided oxygen for breathing after high-altitude ejection. RIGHT: K-1 'Splitshell' with later plexiglass face plate. BELOW: The suit and helmet was tested before each flight for leaks and proper fit.

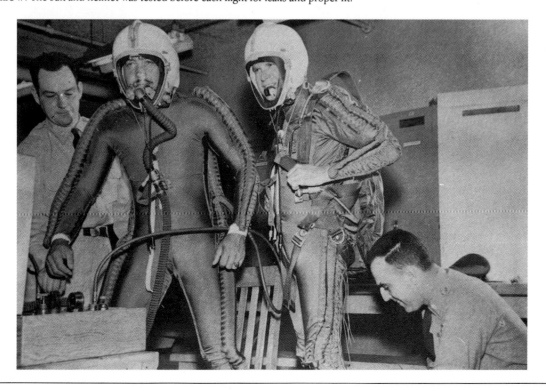

K-1

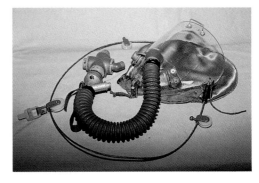

The removable face plate is made of molded plexiglass for light weight, improved visibility and cost reduction. It is electrically heated by way of tiny wires moded within. This face plate uses a low pressure oxygen system. The K-1 was later used with the MA-2 face plate with the high pressure oxygen system. The face plate was stored in a fur lined silk protective bag.

U.S.A.F. K-1 high-altitude partial-pressure helmet. The K-1, made by International Latex Corporation, was the first mass produced U.S. high-altitude helmet in the early 1950s. It was used with the T-1 partial-pressure suit in such aircraft as the B-47, B-36, F-80, F-100 and various X-planes. The U.S.N. also used the K-1 on a limited basis.

The one-piece green fiberglass outer shell connected to a rigid metal frame on the hood.

RIGHT: Standard K-1 microphone. The K-1 also used the MA-2 microphone and was designated as an MB-5 helmet.

CHAPTER 1: U.S.A.F./U.S. ARMY FLIGHT HELMETS & HIGH ALTITUDE HELMETS

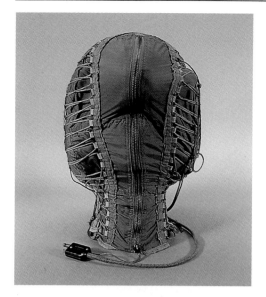

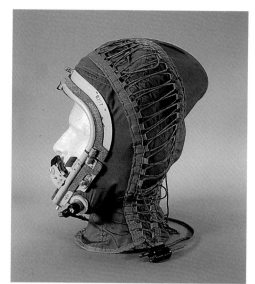

Hood/neck skirt with rigid metal frame.

LEFT: An early version K-1 is shown with external helmet cable mounts. The bailout bottle arrangement is depicted. The pilot is activating the emergency suit pressure and oxygen by way of pulling a green ball nicknamed the 'Green Apple.' Green is the standard indication color for oxygen. RIGHT: The helmet hold-down cable system is clearly depicted.

MA-2

Hood/neck skirt with face plate

MA-2 high-altitude partial-pressure helmet. The MA-2, made by International Latex Corporation, is very similar to the K-1 helmet except that a longer neck skirt is used with additional snaps to connect to the suit. An improved microphone, white fiberglass shell for heat reflection and visibility for rescue, high pressure oxygen system with small diameter reinforced hose and feeding port for water and food during long flights are also incorporated. It was used with the T-1, MC-1, MC-3, MC-3A, MC-4 and MC-4A partial pressure suits. The MA-2 helmet was used by the U.S.A.F., N.A.C.A. and C.I.A. in U-2, F-100, F-101, F-102, F-104, F-106, B-52 and B-58 aircraft as well as several allied air forces such as Sweden, and Great Britain. The helmet was made famous when worn by captured C.I.A. U-2 pilot, Francis Gary Powers. Powers version was unmarked. The MA-2 helmet, introduced in 1956, was used well into the 1960s.

BELOW CENTER: Face plate. BELOW: Face plate in new unissued form with protective visor coating and storage bag.

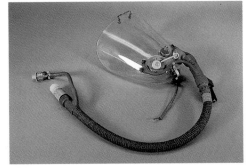
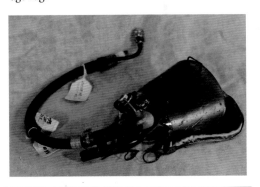

CHAPTER 1: U.S.A.F./U.S. ARMY FLIGHT HELMETS & HIGH ALTITUDE HELMETS

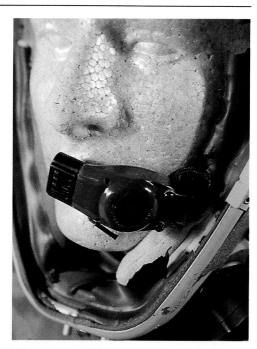

The helmet hold-down cable attachments are inside the shell. RIGHT: Microphone.

LEFT: MA-2 worn by U.S.A.F. Captain J. W. Kittinger, Jr. in 1957 during PROJECT MAN HIGH I high-altitude balloon test to a record 96,000 feet. RIGHT: The feed port allowed the pilot to eat or drink in flight utilizing special tube foods.

JET AGE FLIGHT HELMETS

MA-2 with U.S.A.F. insignia.

CHAPTER 1: U.S.A.F./U.S. ARMY FLIGHT HELMETS & HIGH ALTITUDE HELMETS

MA-1

U.S.A.F. MA-1 high-altitude helmet. The MA-1 helmet, made by Bill Jack Instrument Company in the mid-1950s, was designed to give better visibility then the K-1 or MA-2 but suffered from high weight and oxygen leaks. It could be used with the T-1A and MC-1 type partial-pressure suits or the A/P 22S-2 full-pressure suit. The A/P 22S-2 was the first operational full-pressure suit used by the U.S.A.F.

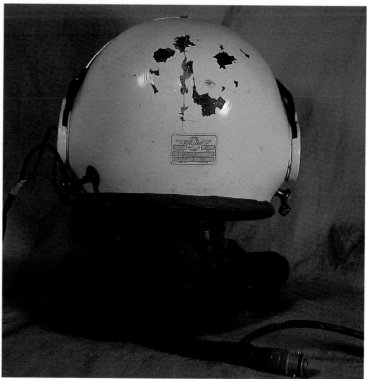

LEFT: The visor is mechanically sealed by way of a top mounted tension bar/cam mechanism. The visor was also electrically heated with vertically running tiny wires molded within. RIGHT: A zippered neck skirt is used.

The same helmet cable restraint system as the K-1/MA-2 is used. One of the two helmet cable attachments is shown.

Early MA-1's used a low pressure oxygen system.

Later MA-1's used a high pressure oxygen system with a smaller diameter oxygen hose.

CHAPTER 1: U.S.A.F./U.S. ARMY FLIGHT HELMETS & HIGH ALTITUDE HELMETS

A/P22S-3 TYPE-1

U.S.A.F. A/P22S-3 Type 1 high-altitude full-pressure helmet. Only one size was available. The silver knob tension system permits the fit of any head by a series of nylon cords connected to the internal suspension system and pads.

The A/P22S-3, first produced in 1960, is the U.S.A.F. version of the B.F. Goodrich U.S.N. Mark IV full-pressure helmet. It is identical except for the silver tension knob, white color and U.S.A.F. decal. The helmet shell is fiberglass and has leather covered ear pads.

The external oxygen regulator incorporates an on/off switch which controls the visor seal.

The clear visor is sealed to the helmet by a pressurized tubular gasket that is fed by bleed-off from the pressure-fed oxygen system.

RIGHT: The tinted outer visor and clear inner visor are actuated by aluminium knobs at the pivot points, one for each visor.

MA-3

U.S.A.F./N.A.S.A. MA-3 high-altitude full-pressure helmet. The MA-3 helmet, made by the Bill Jack Instrument Company, was used in fighter and test aircraft in the early 1960s. It was worn with either the CSU-4/P partial-pressure suit or the A/P22S-2 full-pressure suit. The MA-3 utilized the same cable restraint system as the K-1/MA-2. The tinted visor is mechanically sealed by way of a top mounted tension bar and cam mechanism. Dry oxygen is pressure-fed into the face seal area to defog/defrost the visor.

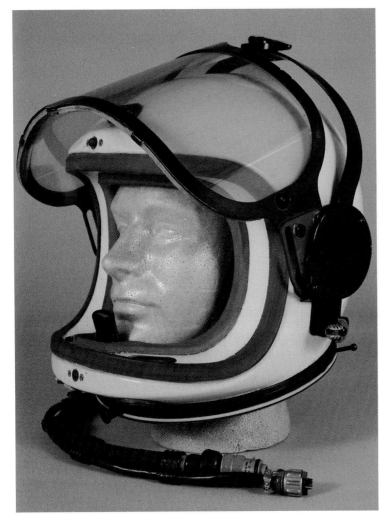 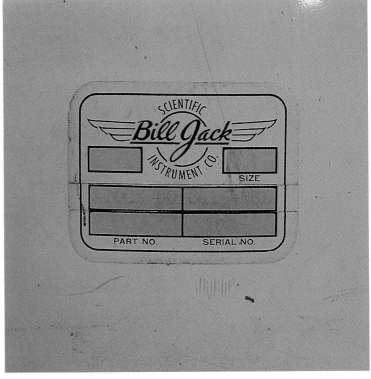

CHAPTER 1: U.S.A.F./U.S. ARMY FLIGHT HELMETS & HIGH ALTITUDE HELMETS

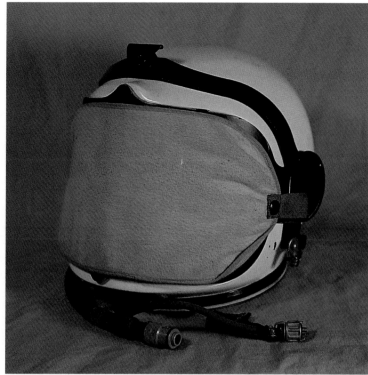

The protective visor cover is used during storage and transportation.

U.S.A.F. pilot Ivan Kinchloe wearing an MA-3 helmet.

XMC-2

The 1958 U.S.A.F. prototype XMC-2 high-altitude full-pressure helmet made by Protection, Inc. A tinted visor was not used. The helmet shell is fiberglass and uses a mechanical visor seal.

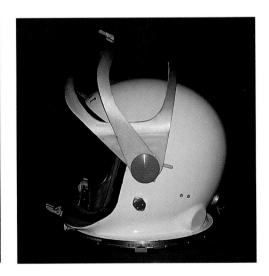

MC-2

The MC-2 full-pressure helmet worn by Joe Walker in 1960 and made by Bill Jack Instrument Company.

The tinted visor is mechanically sealed by way of a front mounted tension bar/cam mechanism. The fiberglass helmet shell is hand made.

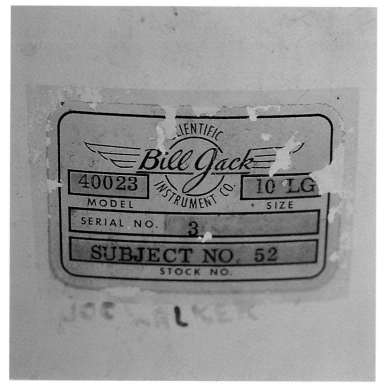

The inner padded helmet is used for comfort and fit only.

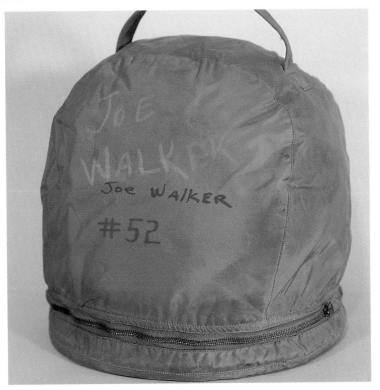

Helmet carry bag.

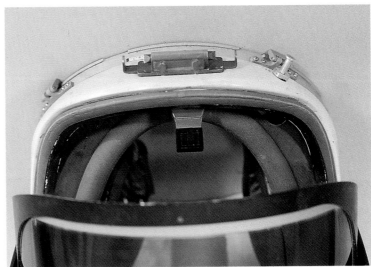

Center mounted microphone.

CHAPTER 1: U.S.A.F./U.S. ARMY FLIGHT HELMETS & HIGH ALTITUDE HELMETS

HGK-13 & HGU-8

HGK-13 full-pressure helmet as used in the X-15 and made by David Clark Company in 1960.

HGU-8 partial-pressure helmet with neck skirt and helmet hold-down cable.

The HGK-13 and HGU-8 helmets are similar in design and were used in the 1960s. Both have two oxygen hoses, one hose supplies breathing oxygen while the other provides 70 psi oxygen to pressurize the inflatable visor seal. The dual oxygen hoses enter the rear of the HGK-13 and internally on the right side of the HGU-8. The communication lead also contains the lead for the electric visor heater for defogging the visor. The HGK-13 did not use the neck skirt and was used with the A/P22S-2 (S-880) full-pressure suit. The HGU-8 used a neck skirt which is tucked into the CSU-4/P partial-pressure suit.

JET AGE FLIGHT HELMETS

HGK-13. Note visor locking knob and anti-suffocation valve.

ABOVE: HGK-13 label. BELOW: HGK-13 internal webbing adjustment take-up knob (right side of oxygen hose connections).

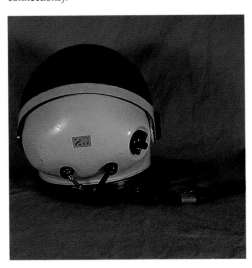

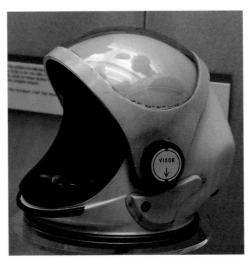

HGK-13 (X-15).

HGU-8.

76

CHAPTER 1: U.S.A.F./U.S. ARMY FLIGHT HELMETS & HIGH ALTITUDE HELMETS

S-901

S-901 high-altitude full-pressure first introduced in 1962 and used in A-12, YF-12A, and SR-71 aircraft.

Visor actuator knob.

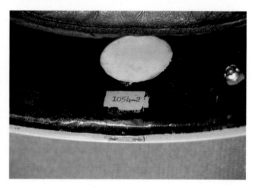

Helmets used by pilots of SR-71 and U-2 aircraft generally do not have a designation, rather they simply go by the manufacturers' suit number, i.e. S-901.

The S-901 helmet uses the HGK-13 shell and has internal dual oxygen hoses.

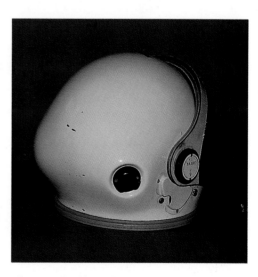

The internal webbing adjustment take-up knob is shown to the left of the tinted visor (shade) actuator knob.

S-1010

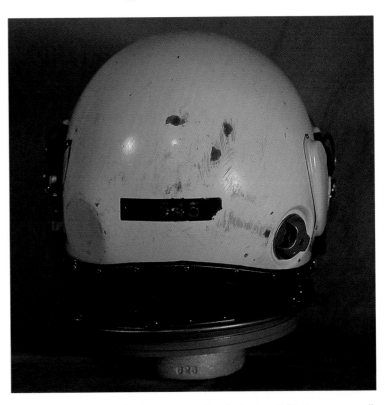

S-1010 high-altitude full-pressure helmet. The S-1010 was designed and produced by the David Clark Company for the U-2C spy plane in the late 1960s. It was eventually replaced by the S-1010B helmet which is the same as the S-1030 helmet.

The anti-suffocation valve is located just below the visor tension bar.

CHAPTER 1: U.S.A.F./U.S. ARMY FLIGHT HELMETS & HIGH ALTITUDE HELMETS

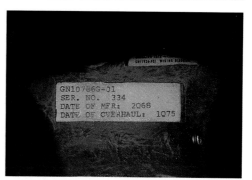

The visor is sealed by way of a mechanical tension bar/cam system with a locking latch in front. The feed port (the hole at the left of the latch) did have a tendency to jam and oxygen would leak from the suit.

A unique combination of articulated neck with hard-mounting neck ring is used to improve flexibility and aid pilot comfort.

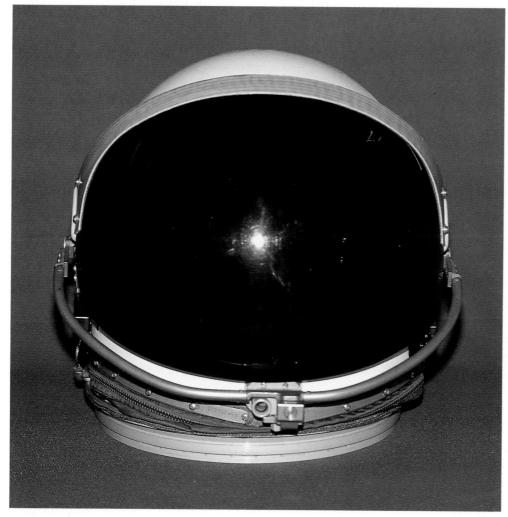

Tinted visor down.

Clear visor.

Tinted and clear visor open.

79

HGK-19

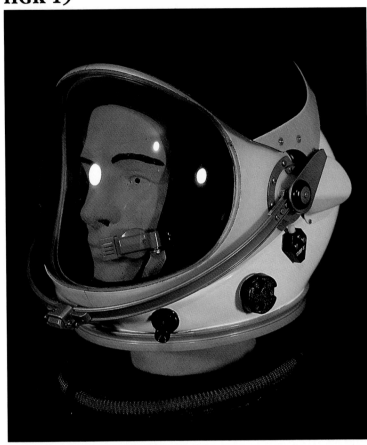
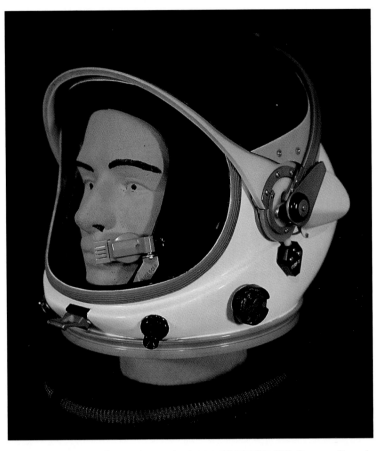

U.S.A.F. HGK-19 high-altitude full-pressure helmet. The HGK-19, first made by the David Clark Company in 1965, was the standard U.S.A.F. high-altitude helmet used into the 1970s. The single oxygen hose HGK-19 is used with the A/P22S-4 and -6 full-pressure suits. The visor is sealed by way of a mechanical tension bar/cam system with a front locking latch.

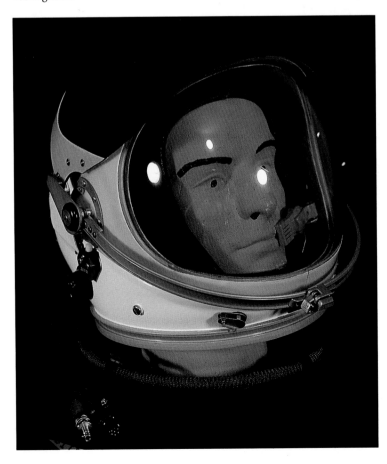
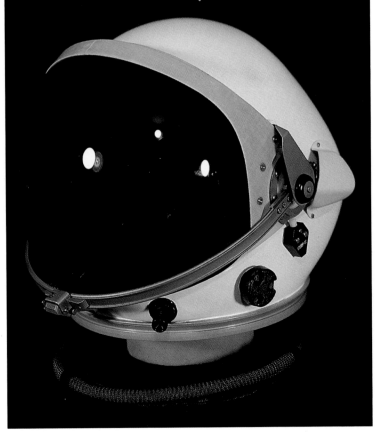

CHAPTER 1: U.S.A.F./U.S. ARMY FLIGHT HELMETS & HIGH ALTITUDE HELMETS

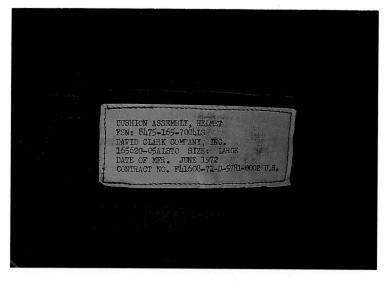

Internal head cushion label.

Internal oxygen regulator.

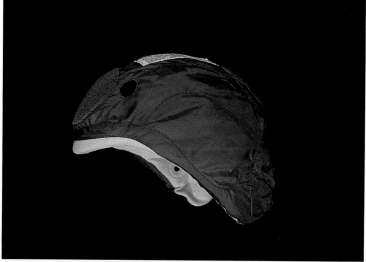

The internal head cushion insert is available in eight sizes and contains the earphone assemblies. The cushion attaches by Velcro to the inside of the helmet.

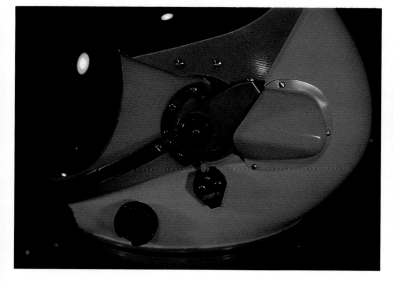

Anti-suffocation valve (lower left).

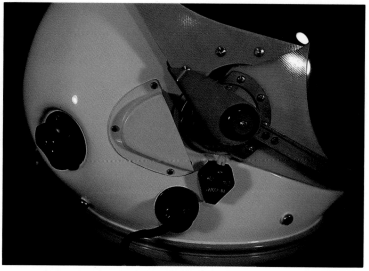

Internal webbing adjustment take-up knob (left).

S-901J

U.S.A.F. S-901J high-altitude full-pressure helmet. The S-901J helmet is similar to the HGK-19 except for the addition of a dual oxygen hose system.

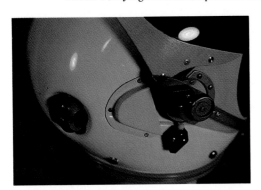

The black switch located behind the visor mount determines which of the two oxygen systems is used.

Dual oxygen hoses enter the rear of helmet.

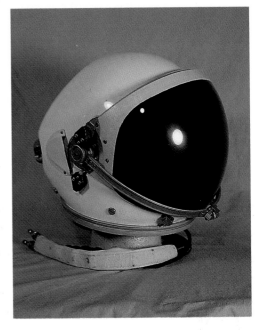

The S-901J helmet is used with the S-970 full-pressure suit (YF-12A) and the S-901J full-pressure suit (SR-71).

S-1030 & S-1031C

U.S.A.F. S-1030 high-altitude full-pressure helmet. The S-1030 helmet was used in the SR-71 'Blackbird' in the late 1970s and is similar to the HGK-19 helmet but with several modifications. The S-1030 has larger internal ventilation ducts, a different internal regulator and visor mechanism. Only one size helmet was produced and used the same head cushion inserts and helmet hold-down system as the HGK-19. The S-1030 replaced the S-1010 helmet. LEFT: The external microphone adjustment knob (below microphone) and anti-suffocation valve (to the right of microphone knob).

The internal webbing adjustment take-up knob is shown on the right side of the helmet.

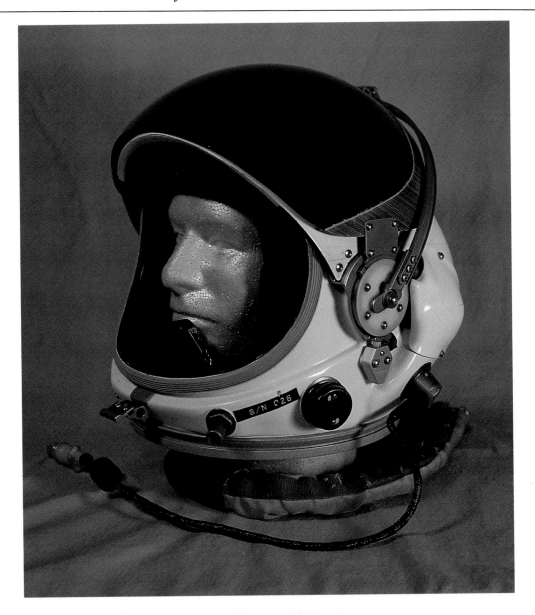

CHAPTER 1: U.S.A.F./U.S. ARMY FLIGHT HELMETS & HIGH ALTITUDE HELMETS

This S-1031C helmet was used in the U-2R/TR-1 in 1991 and is essentially the same as the S-1030 but with no external microphone adjustment. The S1031C has a modified visor tension bar catch.

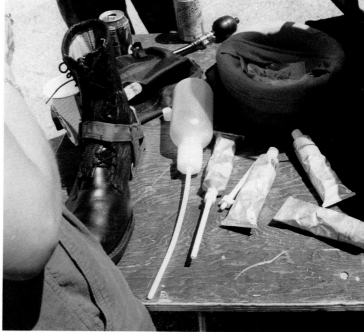

Assortment of food and water available to U-2 and SR-71 pilots.

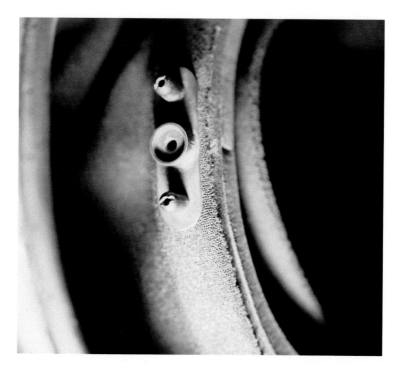

The black face seal has three small screw-holes on the forehead portion for an eyeglass mount.

CHAPTER TWO

U.S.N./U.S.M.C./U.S.C.G. FLIGHT HELMETS & HIGH ALTITUDE HELMETS

H-1

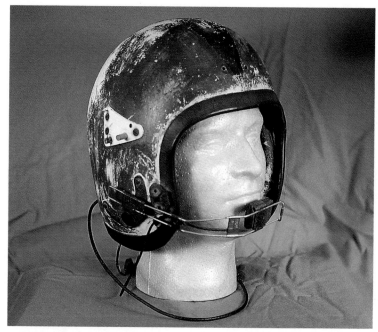
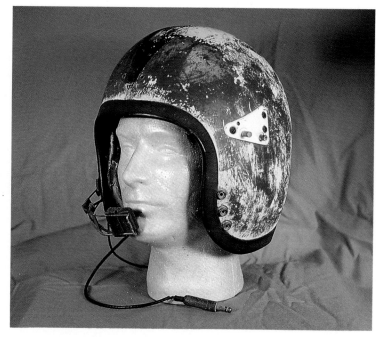

U.S.N. H-1 helmet with boom mounted microphone. The H-1, introduced in 1948 and made by General Textile Corporation, was the first issue U.S.N. hard helmet for fighter aircraft. An A-13A oxygen mask with microphone was used at altitude above 10,000 feet and the boom mounted microphone was used below that altitude. Black rubber goggles were normally worn although the well used H-1 shown had been fitted with a 'P' series external visor at one time. All that remains of that visor is the side brackets and several holes. The H-1 helmet was superseded by the H-3 and H-4 series. Note the unique curved shape at the rear of the helmet shell.

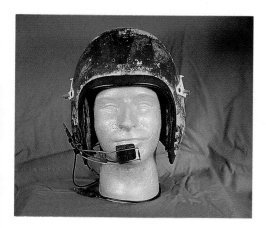
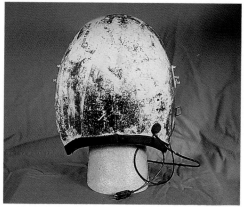
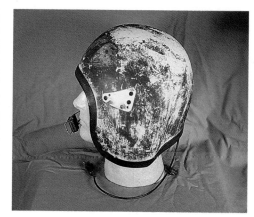

Blue Angels pilot Lcdr. Frank Graham wearing his H-1 helmet.

VF-31 pilots all wearing H-1 helmets (1949).

CHAPTER 2: U.S.N./U.S.M.C./U.S.C.G. FLIGHT HELMETS & HIGH ALTITUDE HELMETS

H-3

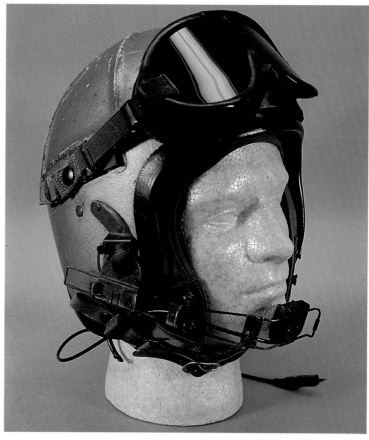

U.S.N. H-3 helmet with boom mounted M-6A/UR microphone and U.S.N. goggles. The H-3 was a standard U.S.N fighter aircraft helmet used during and after Korea. It uses a cloth inner helmet secured to the rigid protective shell by 'pull the dot' snap fasteners on the cheek flaps. The gold color was standard issue on many U.S.N. helmets of the 1950s and early 1960s but many will be found with squadron and personal paint schemes. The rigid shell is constructed of a fiberglass cloth reinforced with epoxy resin.

The goggles are attached to snaps on the helmet adjustment straps.

The H-3 helmet reinforcing ridges differ from that of the later H-4 helmet.

89

JET AGE FLIGHT HELMETS

The inner cloth helmet incorporates the earphones and can be used in-dependently of the outer rigid helmet.

Outer helmet shell label.

Inner cloth helmet label.

RIGHT: Black rubber U.S.N. goggles manufactured by Rochester Optical.

H-4

U.S.N. H-4 helmet with boom mounted M-6A/UR microphone, U.S.N. goggles and MS22001 oxygen mask. The H-4 helmet was used by the U.S.N. and U.S.M.C. through the 1950s and into the early 1960s. Canada and other foreign countries also used variations of the H-4.

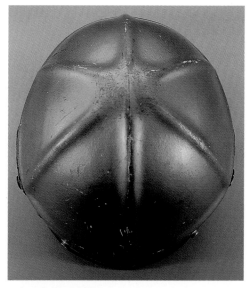

H-4 reinforcing ridges.

ABOVE AND BELOW: Cloth inner helmet.

FAR LEFT: MC-3A oxygen connector.

LEFT: Inner cloth helmet label.

Outer helmet shell label.

Label attached to suspension webbing on outer helmet shell.

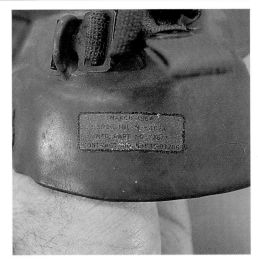
Oxygen mask label.

U.S.N pilot wearing his H-4 helmet and goggles.

CHAPTER 2: U.S.N./U.S.M.C./U.S.C.G. FLIGHT HELMETS & HIGH ALTITUDE HELMETS

 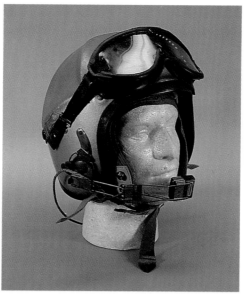 

The goggle attachments snaps are located on the helmet shell below the adjustment webbing.

A U.S.N. Banshee pilot in 1953 wearing an H-4 modified with a 'P' helmet external visor. Note the two extra retaining straps in front securing the inner helmet to the rigid helmet. These additional straps (along with two added in the back) were a modification to better secure the rigid helmet to the inner cloth helmet. It was found that without this modification, the rigid helmet would be blown off of the pilot's head during ejection.

SPH-1

U.S.N. SPH-1 helicopter helmet with M-6A/UR microphone mounted on chin cup. The SPH-1, made by International Latex Corp., was a primary U.S.N. helicopter helmet in the early 1950s. Its shell is constructed of a fiberglass cloth reinforced with epoxy resin. The rows of snaps on each side of the helmet could accommodate an oxygen mask. Sunglasses could have also been worn for eye protection.

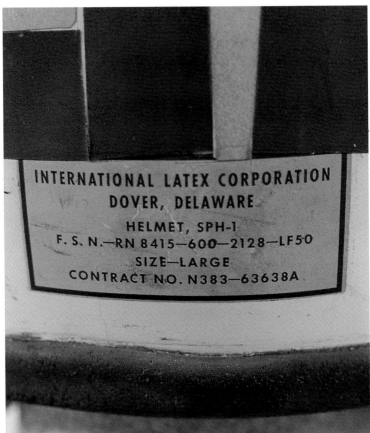

SPH-2

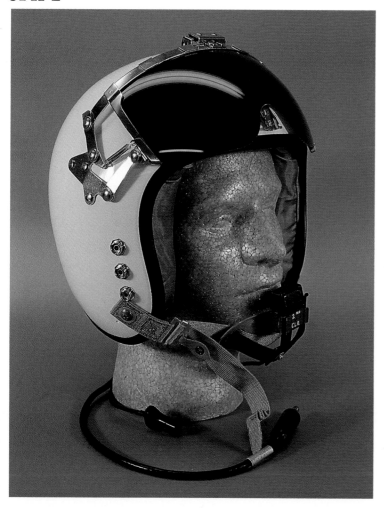
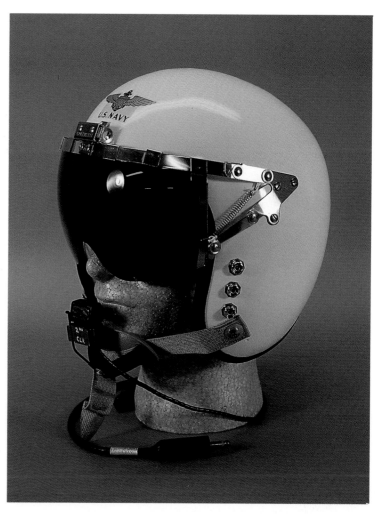

U.S.N. SPH-2 helicopter helmet with external neutral grey visor and CW-292/U microphone mounted on chin cup. The SPH-2, made by CaL-Mil Plastic Products, Inc., was used in the 1950s and early 1960s. An oxygen mask could be attached to the snaps on each side of the helmet and the visor was locked 'up' or 'down' by a center squeeze type locking mechanism.

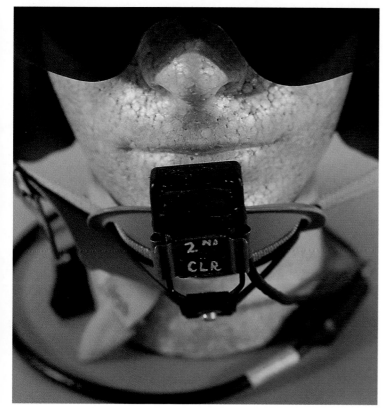
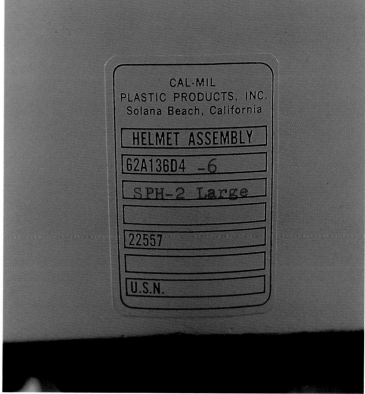

BPH-2

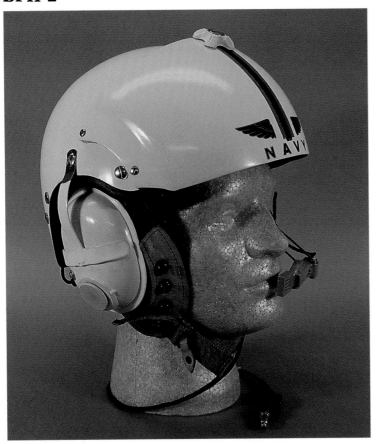
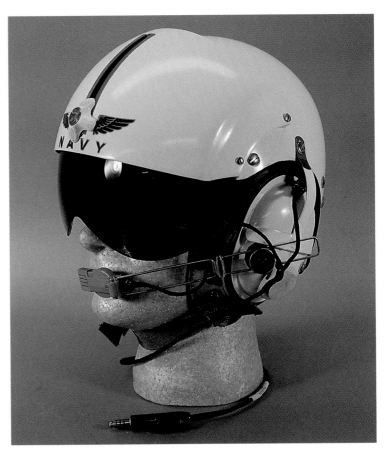

U.S.N. BPH-2 patrol aircraft helmet with neutral grey visor, boom mounted M-87/A1C microphone and H87B receivers. The Gentex BPH-2 helmet was used in the mid 1960s, early 1970s in reconnaissance and patrol aircraft, and also used without a visor system for crew and flight deck use. The shell is constructed of a fiberglass cloth reinforced with epoxy resin. A clear visor was available.

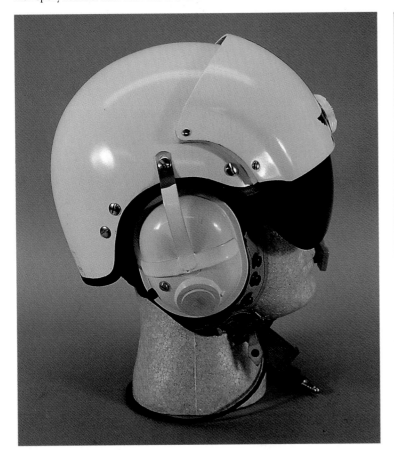
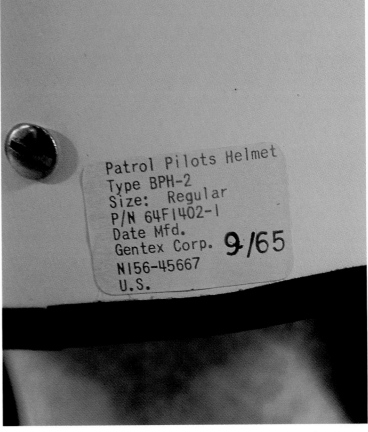

CHAPTER 2: U.S.N./U.S.M.C./U.S.C.G. FLIGHT HELMETS & HIGH ALTITUDE HELMETS

APH-5

U.S.N. APH-5 helmet with neutral grey visor and MS22001 oxygen mask. The Gentex APH-5 helmet was used during the 1950s and 1960s in various fighter aircraft. Its shell is constructed of a fiberglass cloth reinforced with epoxy resin. BELOW RIGHT: An APH-5 modified with 'Hardman' oxygen mask receivers and mask mounted mini oxygen regulator.

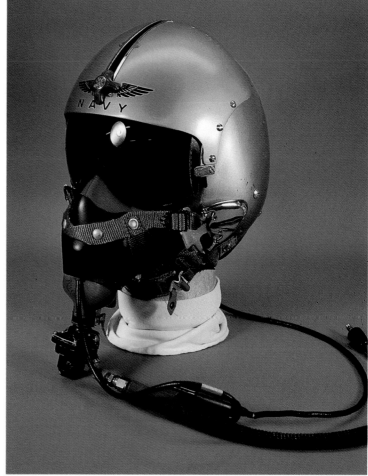

JET AGE FLIGHT HELMETS

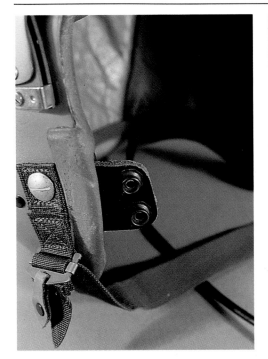

The oxygen mask is attached to snaps on leather tabs on earlier APH-5's.

The MS22001 oxygen mask (The Navy also uses the A-13A designation for this mask).

Oxygen mask with manufacture date stamp.

ABOVE: 'Hardman' receiver. BELOW: Early APH-5's used this round visor knob incorporating a push button locking device.

ABOVE: Later type visor locking knob.

APH-6A

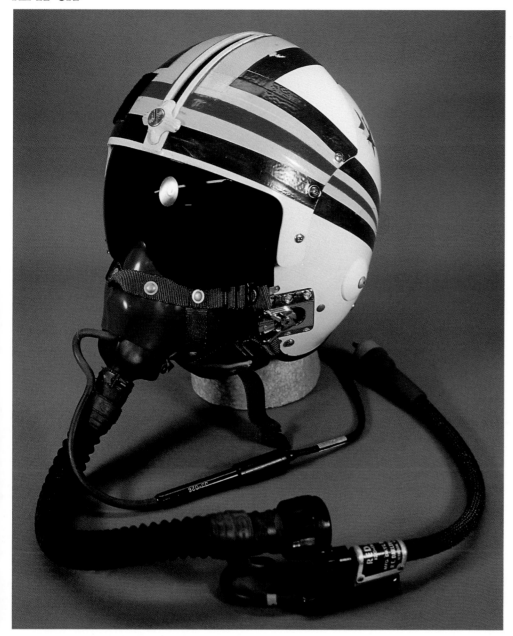
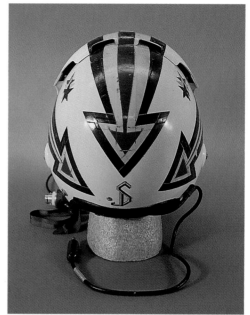

U.S.N. APH-6A single visor helmet with neutral grey visor and A13A (MS22001) oxygen mask with REDAR oxygen hose. The APH-6A helmet was used from the 1960s through the 1980s in fighter, bomber and a variety of other aircraft by the U.S.N., U.S.M.C. and U.S.C.G. Boom microphones are used in many non-fighter type aircraft. The shell is constructed of a polyester impregnated fiberglass cloth laminate. The APH-6D designation is a similar helmet with the exception of the liner and avionics. RIGHT: The black patch of Velcro on the visor housing is used to attach a distress strobe or light in the event of bailout or ditching.

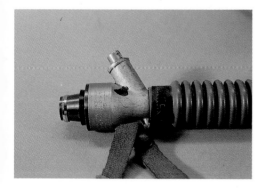

MC-3A oxygen connector.   The unique 'Butterfly' retention and release assembly used to secure the oxygen mask.

JET AGE FLIGHT HELMETS

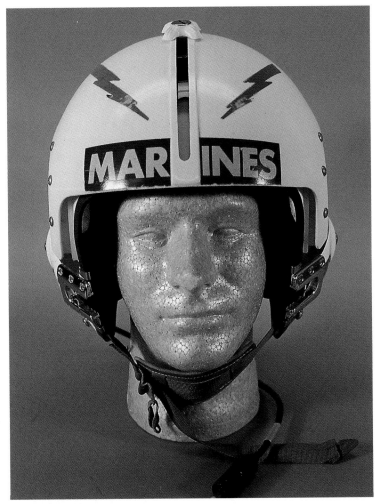

U.S.M.C. APH-6A.

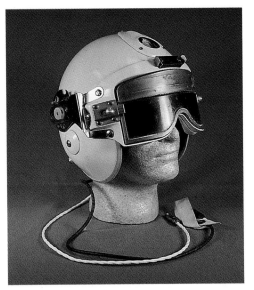
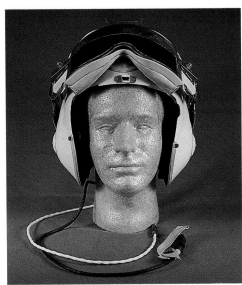
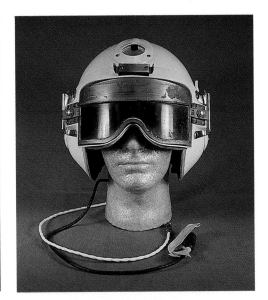

ABOVE LEFT: This APH-6 helmet shell has been fitted with a nuclear flash visor. CENTER: The visor has been designated a DH-101A. RIGHT: The helmet has not yet been fitted with a chin snap or oxygen receivers. The original lens are automatically darkened after a nuclear flash and can be replaced with a clear lens. No other information is currently available about this helmet and visor.

CHAPTER 2: U.S.N./U.S.M.C./U.S.C.G. FLIGHT HELMETS & HIGH ALTITUDE HELMETS

APH-6A (VTAS I)

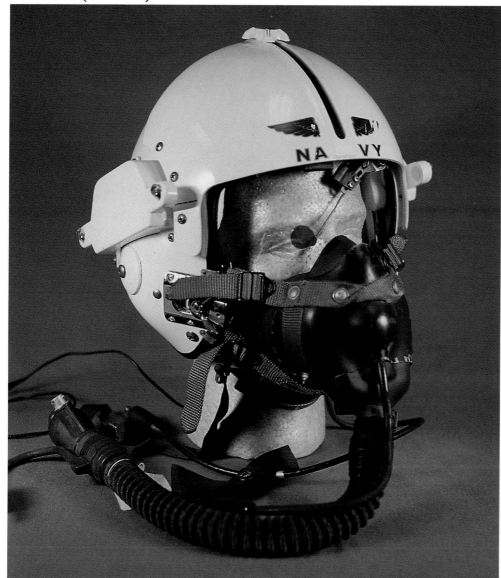
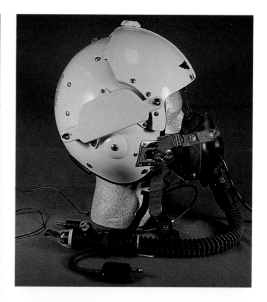
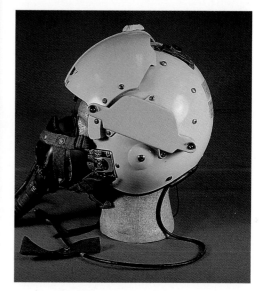

U.S.N. APH-6A incorporating the Visual Target Acquisition System (VTAS I) and A13A (MS22001) oxygen mask. The VTAS I was the first helmet sighting system introduced for operational fighters and was adopted in 1972. This system used a 'Granny glass' sight piece that was deployed and stowed inside the visor using a slide knob located on the helmet shell. The sensor assemblies are separated from the sight mechanism and are attached to either side of the helmet. The 'Granny Glass' sight piece was not successful with pilots. The exit pupil of the sight piece mechanism was too small and any shifting of the helmet under 'G' loading caused the sight piece glass to shift in front of the pilot's eye, causing him to lose sight of the reticle image. This problem was rectified with the visor reticle helmet mounted unit used in the VTAS II system. BELOW LEFT: Sight piece slide knob. CENTER: Sight piece in position. RIGHT: MC-3A oxygen connector.

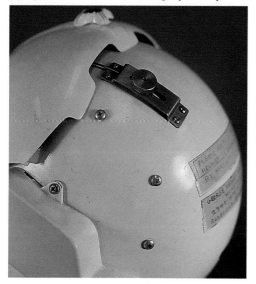

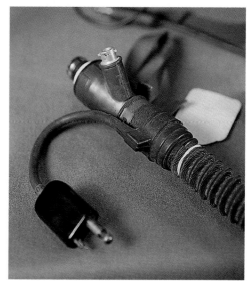

APH-6B

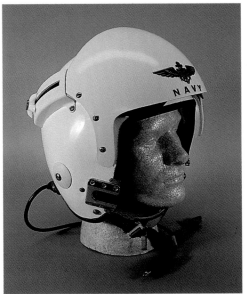

U.S.N. APH-6B dual visor helmet with A13A (MS22001) oxygen mask. The APH-6B helmets were used in such aircraft as the Intruder fighter/bomber and also by the U.S.M.C. The APH-6B uses the same shell as the single visor APH-6A. The APH-6C dual visor helmet is similar to the APH-6B except for the liner and avionics. This helmet was made by Sierra Engineering Corporation.

RIGHT: This helmet uses a three piece visor housing assembly with separate locking knob tracks.

FAR RIGHT: This black APH-6B uses a one piece visor housing assembly.

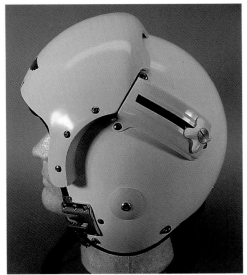
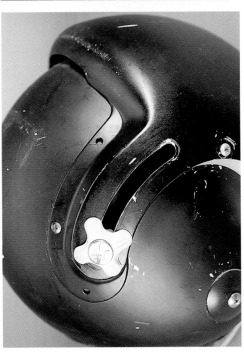

CHAPTER 2: U.S.N./U.S.M.C./U.S.C.G. FLIGHT HELMETS & HIGH ALTITUDE HELMETS

HGU-20/P

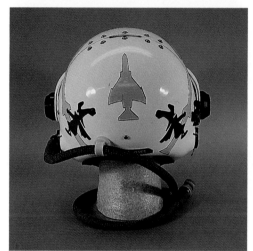

U.S.N. HGU-20/P 'Clam Shell' helmet with clear visor, neutral grey visor and M-94A/A microphone. The U.S.N. HGU-20/P 'Clam Shell' helmet, made by Robert Shaw Controls Company, was flight tested in fighter and fighter bomber aircraft from the mid 1960s until 1971. It was designed to eliminate the oxygen mask and to prevent the helmet from being blown off during ejection. Oxygen was made available through a regulator and a series of small holes in ductwork surrounding the face. The U.S.A.F. also flight tested this helmet and used it in the B-1A program.

Oxygen regulator within the helmet.

Clear visor locking knob, shell locking mechanism and internal webbing adjustment knob.

HGU-20/P pilot helmet.

JET AGE FLIGHT HELMETS

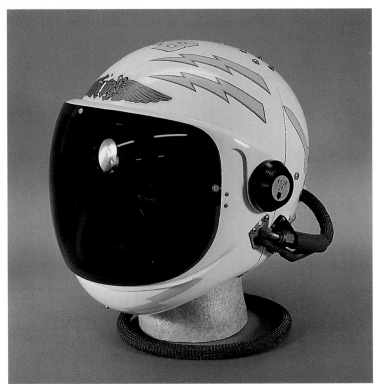

HGU-20/P helmet used by Weapons Officer in F-4 fighter. Note the wing decal design.

Donning the helmet requires opening the helmet shell, positioning the wearer's head between the two halves and then closing and locking the two halves, hence the nickname 'Clam Shell.'

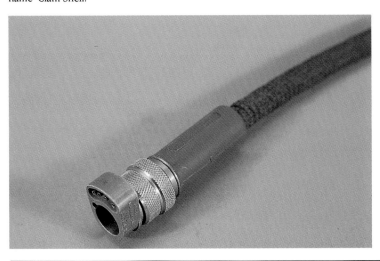

LEFT: The REDAR hose connection incorporates both the oxygen and communication systems.

CHAPTER 2: U.S.N./U.S.M.C./U.S.C.G. FLIGHT HELMETS & HIGH ALTITUDE HELMETS

This heavily modified HGU-20/P was a limited production chemical/biological protection helmet and was designated as an HGU-15/P.

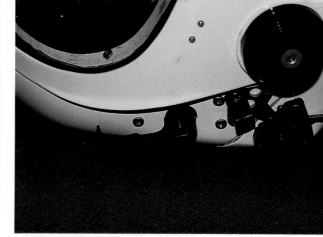

A special mask is used for oxygen.

The lever activates the visor seal.

105

SPH-3B

U.S.N. SPH-3B dual visor helicopter helmet with boom mounted M-87/A1C microphone. The SPH-3B has also been used by the U.S.C.G. and U.S.M.C. since the 1970s and is still in use today. Canada (SPH-3CF) and Germany (SPH-4G) also use this helmet. A neutral grey and clear visor are provided and its shell is constructed of fiberglass cloth impregnated with epoxy resin. The SPH-3C helmet is similar except that the shell is made of five plies of Kevlar fabric reinforced with epoxy resin.

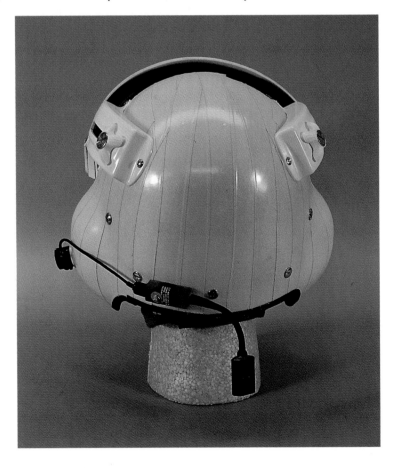

HGU-34/P

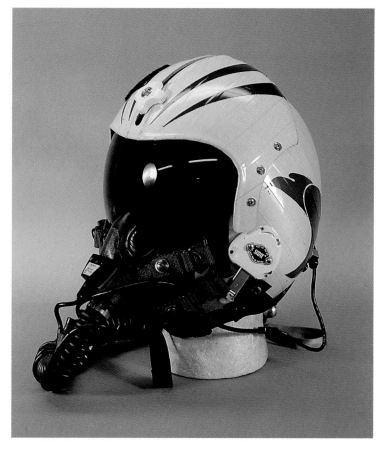

U.S.N. HGU-34/P helmet with neutral grey visor and MBU-14/P oxygen mask. The HGU-34/P helmet has been in use in fighter and other aircraft since the mid 1970s by the U.S.N. and U.S.M.C. Its shell is made of fiberglass cloth reinforced with epoxy resin. Clear visors were also used with this helmet. The HGU-33/P and 34/P are identical helmets except for the inner liners. The HGU-33/P uses a form fit liner v.s. a styrofoam liner in the 34/P. These helmets were made by Gentex corporation and Sierra Engineering Company. This helmet is comprised of a PRK-37/P shell assembly and PRK-40/P foam liner assembly. Many helmet variations exist using the PRK-37/P shell. The U.S.N. assigns a specific HGU nomenclature depending upon the type of aircraft and helmet assembly.

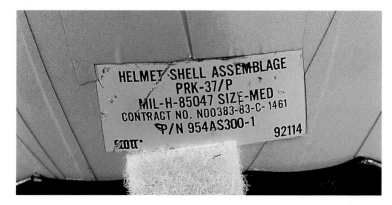

ABOVE: Helmet shell label. BELOW: Helmet liner label.

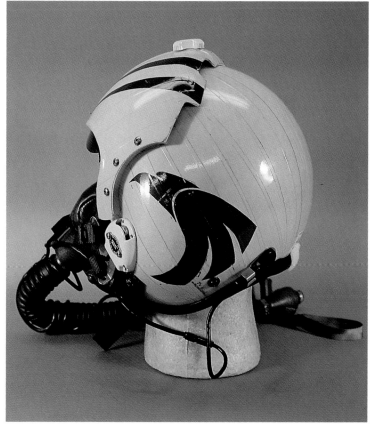

JET AGE FLIGHT HELMETS

Removable camouflage cloth cover.

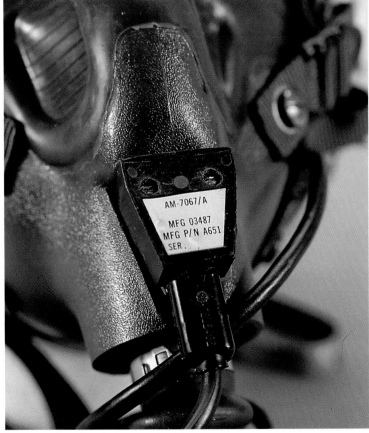

ABOVE LEFT: Offset bayonet and oxygen suspension system. RIGHT: MBU-17(V)1/P oxygen mask. This green mask is an early MBU-12/P with the addition of a special U.S.N. communication system. The Navy redesignates the MBU-12/P mask as a -14/P, -15/P, -16/P, -17/P depending upon the type of communication setup.

CHAPTER 2: U.S.N./U.S.M.C./U.S.C.G. FLIGHT HELMETS & HIGH ALTITUDE HELMETS

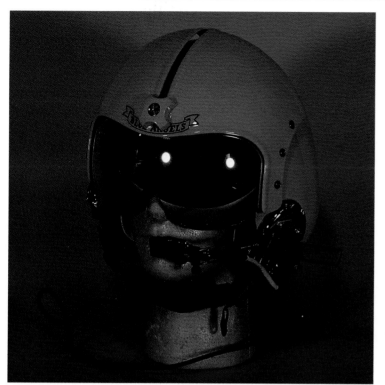
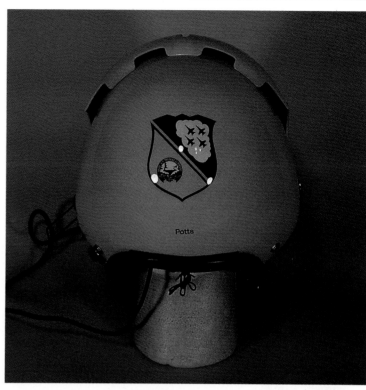

U.S.N. Blue Angels HGU-33/P with gold tinted visor and boom mounted microphone.

HGU-45/P

U.S.M.C. HGU-45/P dual visor helmet with boom mounted M-87/A1C microphone as used in an OV-10 Bronco in the late 1980s and early 1990s. A neutral gray visor (shown) and clear visor are used.

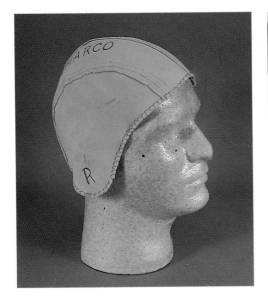

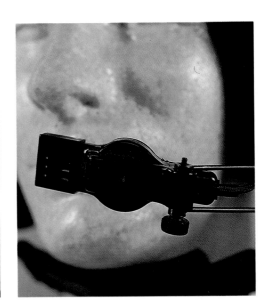

Skull caps have been worn under most flight helmets since the 1970s for comfort and absorption of perspiration.

CHAPTER 2: U.S.N./U.S.M.C./U.S.C.G. FLIGHT HELMETS & HIGH ALTITUDE HELMETS

HGU-46/P (VTAS II)

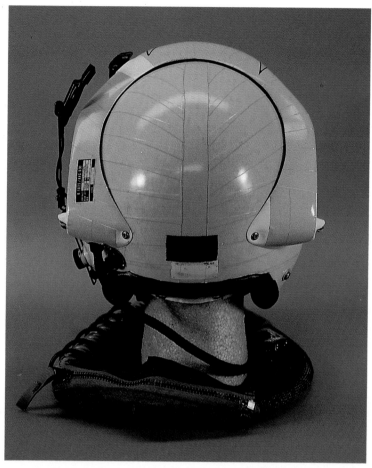

U.S.N. HGU-46/P helmet incorporating the R-1825/AVG-8(V) Visual Target Acquisition System (VTAS II), MBU-14(V)1/P oxygen mask and boom mounted M-87-AIC microphone. The VTAS II (an improved version of the VTAS I) is a head position referencing system or 'Helmet Sight' system. This helmet was used in F-4 Phantom aircraft starting in the mid 1970s and slaved the aircraft radar or sidewinder missile to the pilot's line-of-sight for use in air-to-air combat. The pilot would superimpose a collimated reticle image upon a target aircraft, actuate a trigger switch on the control stick and the radar or missile would lock onto the target. The pilot would then use his conventional on-board fire control system.

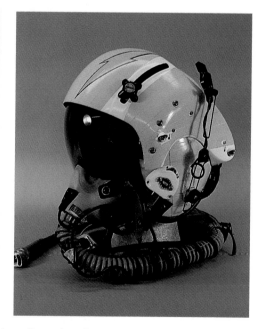

The VTAS II uses a form fit PRK-37/P shell. The parabolic visor provides the pilot with a collimated reticle image.

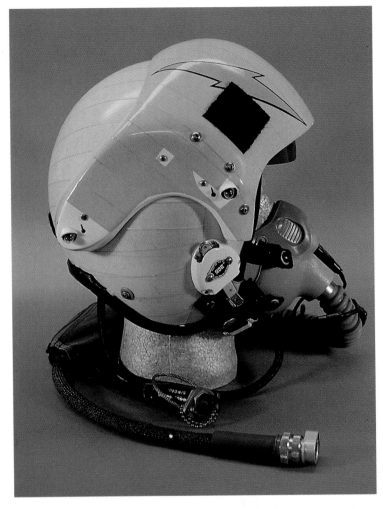

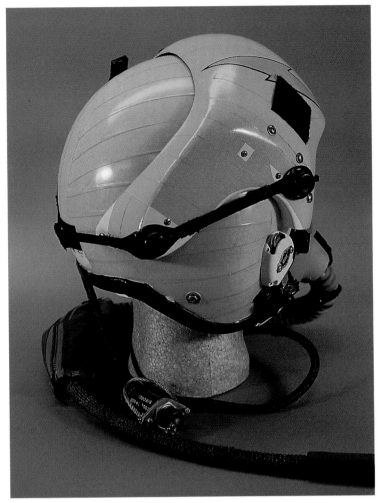

Two sensor electronic assemblies, one mounted on each side of the receiver housing, assist in converting the pilot's line-of-sight into aircraft coordinates.

Sensor protection caps are used in transport and storage.

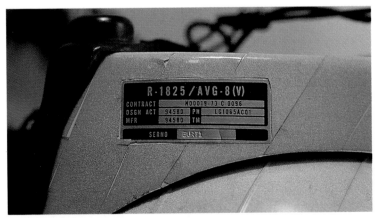

ABOVE: VTAS II electrical connection.

RIGHT: The REDAR (R.E. Darling Co., Inc). connection is enclosed in a zippered bag that attaches to the survival vest.

CHAPTER 2: U.S.N./U.S.M.C./U.S.C.G. FLIGHT HELMETS & HIGH ALTITUDE HELMETS

SPH-5CG

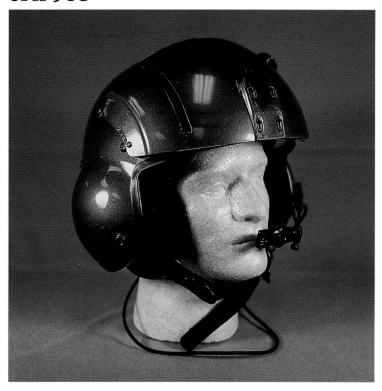
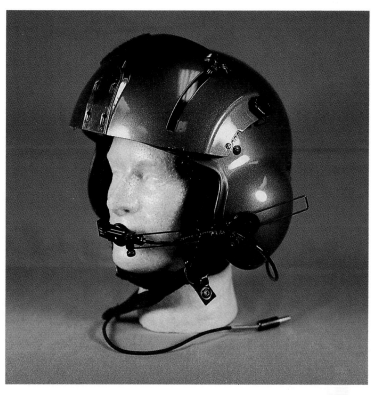

Current issue Gentex U.S.C.G. SPH-5CG dual visor helicopter helmet with boom mounted M-87/A1C microphone and quick disconnect mounting bracket for the ANVIS-6 night vision goggles. The helmet shell is lightweight composite reinforced with epoxy resin. A neutral grey and clear visor are provided. The chin strap is secured using 'D' rings and snaps.

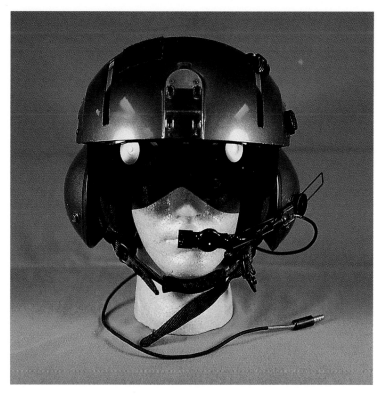

HGU-67/P

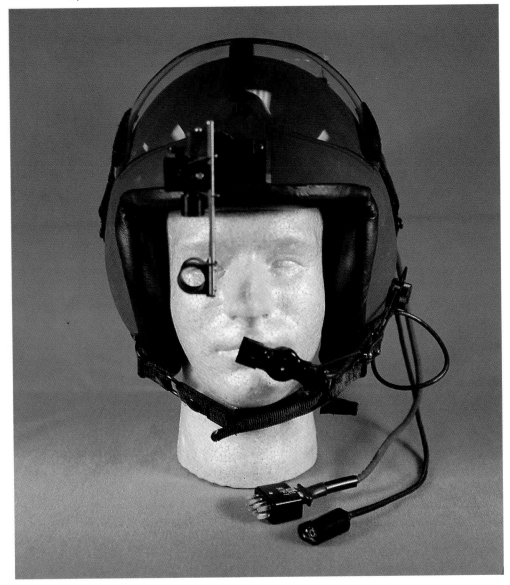

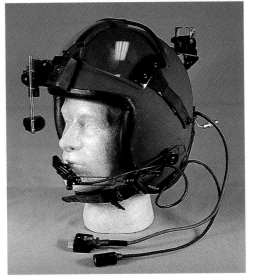

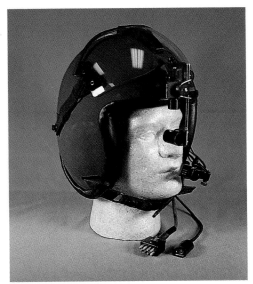

Current issue U.S.M.C. HGU-67/P 'Advanced Cobra' helmet with external clear visor, boom mounted M-87/A1C microphone and helmet sighting assembly. The Gentex HGU-67/P helmet replaces the SPH-3C and its shell is made of a laminated graphite and ballistic nylon. The chin strap is a one piece assembly integrated with the nape strap and threaded through the helmet. Cross straps threaded through the nape strap pad adjust simultaneously with the chin strap to provide fit. A chamois backed leather cover (not shown) protects the visor during transport and storage.

A neutral grey visor can also be used and is located on top of the clear visor. The additional visor strap attaches to the snap located on the clear visor strap.

Sight electrical interface connection.

Sight mounting bracket.

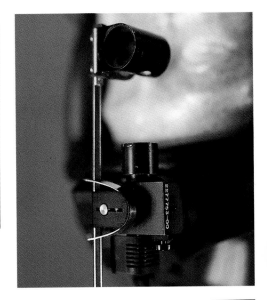

CHAPTER 2: U.S.N./U.S.M.C./U.S.C.G. FLIGHT HELMETS & HIGH ALTITUDE HELMETS

HGU-85/P

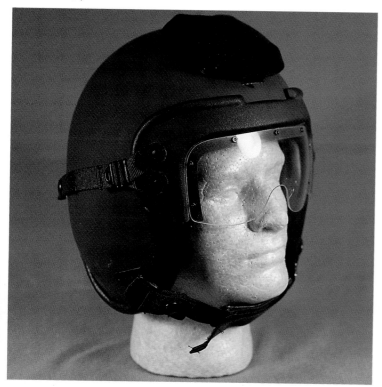
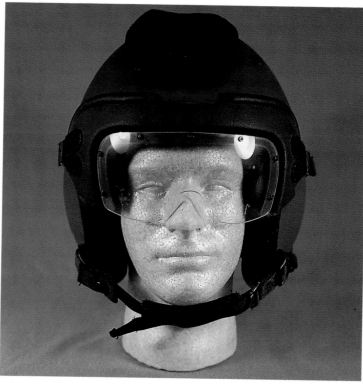

HGU-85/P 'Night Attack' helmet with NVG mounting plate, mounting plate protective cover and clear visor. The Gentex HGU-85/P 'Night Attack' helmet is an interim U.S.N./U.S.M.C. design to meet the requirements of the tactical night operating environment using night vision goggles (NVG). A neutral grey visor is used for daytime operation and the clear visor for night operations. A contoured black cloth lens cover (not shown) is available. Lightweight bayonet receivers can be provided to accommodate the MBU-12/P oxygen mask. The one piece integrated chin/nape strap is similar in design to the HGU-67/P helmet. This helmet and the HGU-55/P will be replaced in 1996 by the U.S.N. 'Combat Edge' HGU-68/P helmet. The HGU-85/P shell is a composite made of pressure molded, laminated graphite and ballistic nylon.

HGU-68/P

U.S.N. HGU-68/P 'Combat Edge' helmet with neutral grey visor and MBU-20/P oxygen mask. The U.S.N. HGU-68/P 'Combat Edge' helmet, made by Gentex Corporation, is part of a tactical total pressure breathing system being introduced in 1996 to replace the HGU-55/P helmet. This system is the U.S.N. configuration of the U.S.A.F. 'Combat Edge' system and will be used in F-18 aircraft. The helmet will be used in conjunction with a low profile MBU-20/P side hose oxygen mask, CSU-17P counter pressure vest, chest mounted regulator and full coverage ankle-to-abdomen lower 'G' garment. BELOW RIGHT: Chest mounted regulator.

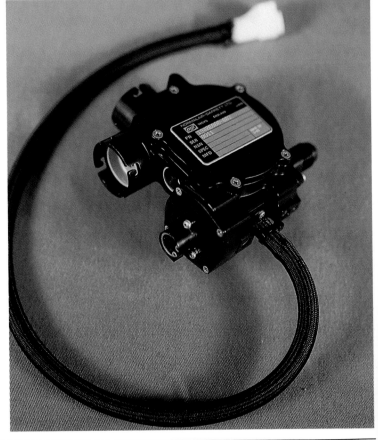

CHAPTER 2: U.S.N./U.S.M.C./U.S.C.G. FLIGHT HELMETS & HIGH ALTITUDE HELMETS

The external visor does not utilize a visor housing and slides on tracks. A locking knob secures the visor in the desired position.

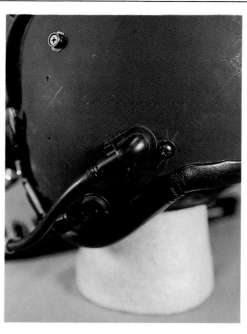

Quick disconnect helmet bladder supply hose from MBU-20/P oxygen mask.

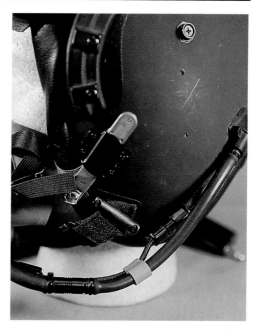

One piece integrated chin/nape strap with strap 'locking bar.'

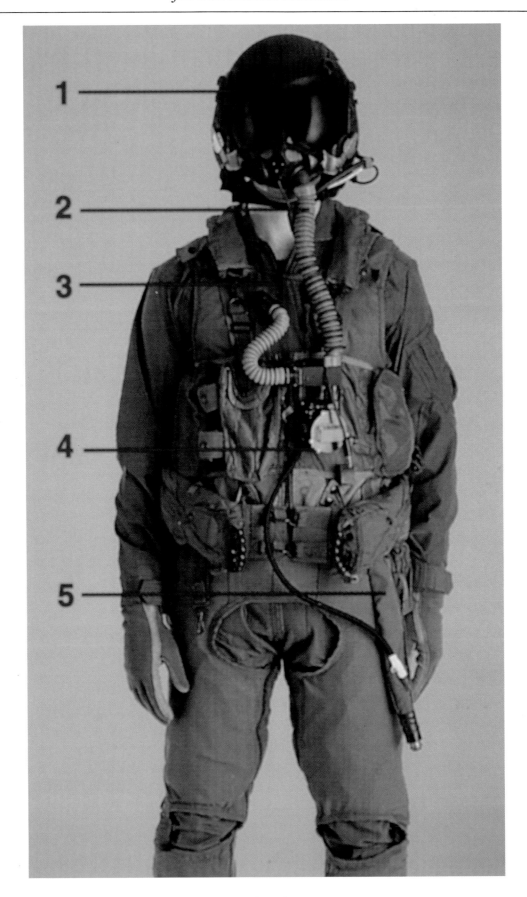

U.S.N. 'Combat Edge' system

1. HGU-68/P helmet
2. MBU-20/P oxygen mask
3. CSU-17/P counter pressure vest
4. Chest mounted regulator
5. Lower 'G' garment

AIR COMBAT SYSTEM

'Air Combat System' dual visor helmet with an MBU-20/P oxygen mask incorporating lightweight bayonets and receivers. The Gentex 'Air Combat System' is a fully integrated support system for tactical aircraft similar to the U.S.N. and U.S.A.F. 'Combat Edge' system. The system consists of a dual visor helmet (No designation has been assigned), MBU-20/P oxygen mask, two-piece combined flight suit/anti-G suit, chest counter pressure garment and chest-mounted regulator. The system has been flown in YF-22, YF-23, F-15 and F-16 aircraft and is currently only used for F-18 export aircraft. It has not been adopted by the U.S. military at this time. The dual visor assembly utilizes an infinitely adjustable (full up to full down) rotary visor mechanism. The left side mechanism controls the inner clear visor and the right side mechanism controls the outer sunshade visor. The helmet shell is constructed from an advanced lightweight Kevlar. BELOW RIGHT: Protective visor cover installed.

CHAPTER 2: U.S.N./U.S.M.C./U.S.C.G. FLIGHT HELMETS & HIGH ALTITUDE HELMETS

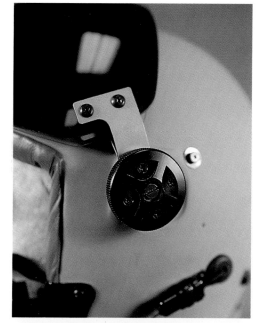

LEFT: The lightweight mask bayonet receivers provide a direct pneumatic coupling to the automatic mask tensioning bladder situated at the rear of the helmet.

HIGH ALTITUDE HELMETS
(PARTIAL-PRESSURE AND FULL-PRESSURE)

MK-I

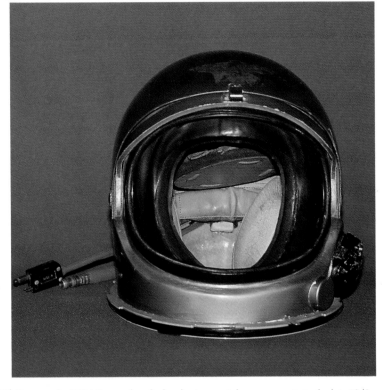

A U.S.N. MK-I, Model 3 high-altitude full-pressure helmet made by B.F. Goodrich in 1956. Because the U.S.A.F. was already developing partial-pressure suits in the late 1940s, the U.S.N. was tasked with developing full-pressure suits. The MK-I was the very first full-pressure helmet to actually be used in flight.

The clear visor retracts into a built-in visor housing. The internal webbing adjustment knob is on the left. A tinted visor was not used.

CHAPTER 2: U.S.N./U.S.M.C./U.S.C.G. FLIGHT HELMETS & HIGH ALTITUDE HELMETS

The clear visor is sealed by an inflatable face seal.

All U.S.N. full-pressure helmets use an external oxygen regulator.

MK-II

A U.S.N. MK-II high-altitude full-pressure helmet made by B.F. Goodrich in 1957. The MK-II helmet offered improved visibility and comfort over the MK-I. The clear visor retracts into a built-in visor housing. The gold color internal webbing adjustment knob is located at the right rear. A tinted visor was not used.

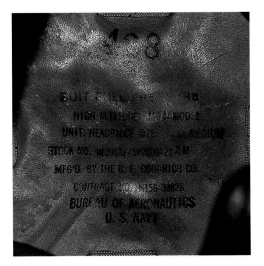

External oxygen regulator with on/off switch.

Microphone and adjustment knob. The visor is sealed by an inflatable face seal.

CHAPTER 2: U.S.N./U.S.M.C./U.S.C.G. FLIGHT HELMETS & HIGH ALTITUDE HELMETS

MK-III

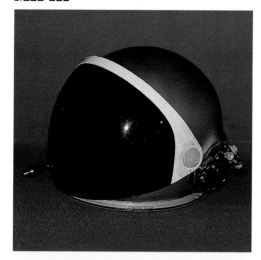
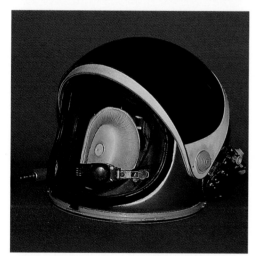

A U.S.N. MK-III high-altitude full-pressure helmet made by B.F. Goodrich in 1958. The MK-III was the first U.S.N. full-pressure helmet to use a clear and tinted visor. Note the external oxygen regulator with on/off switch.

Microphone and adjustment knob. The visor is sealed by an inflatable face seal.

The internal webbing adjustment knob is located at the right rear.

U.S.N. pilot wearing a MK-III helmet and full-pressure suit undergoing ejection seat training.

CHAPTER 2: U.S.N./U.S.M.C./U.S.C.G. FLIGHT HELMETS & HIGH ALTITUDE HELMETS

MK-IV

A U.S.N. MK-IV high-altitude full-pressure helmet. The MK-IV, first made by B. F. Goodrich in 1959, was the standard U.S.N. high-altitude helmet and remained in service into the early 1970s. It was also the basis for the Project Mercury space helmet. The MK-IV uses an inflatable face seal to seal the clear visor.

External oxygen regulator with on/off switch.

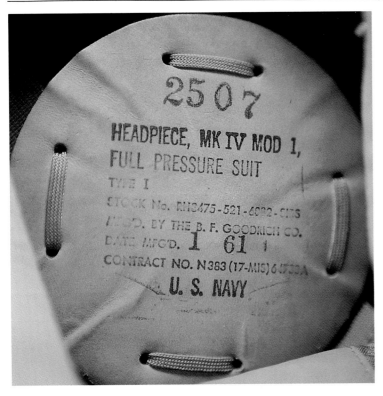

Plastic carrying/storage container.

CHAPTER THREE

FOREIGN FLIGHT HELMETS

CANADA - H4-1

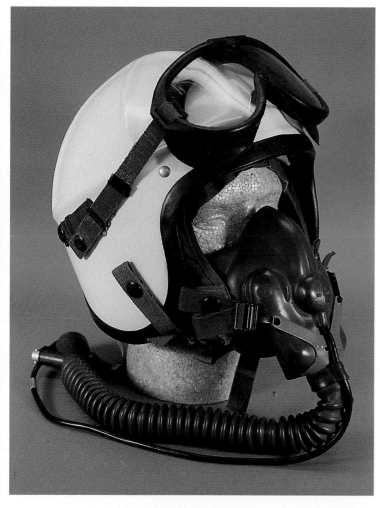

The H4-1 was first issued in 1954. Produced by Gentex, the H4-1 was used in both fighters and helicopters.

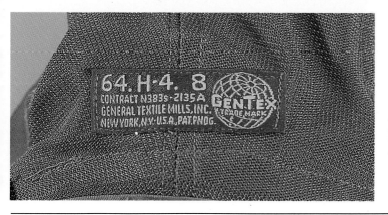
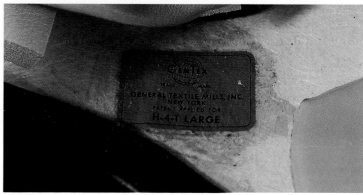

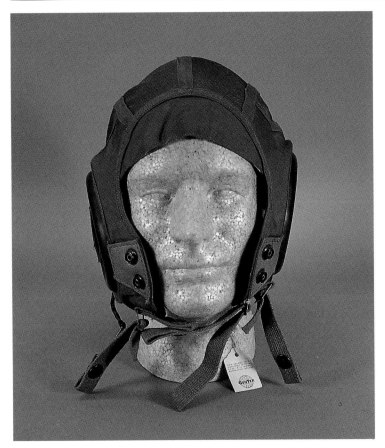 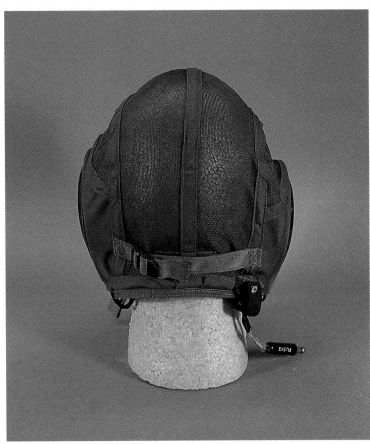

The inner helmet of the H4-1 has an adjustable nap strap.

RIGHT: A type MS22001 oxygen mask is standard.

CANADA - 41-1

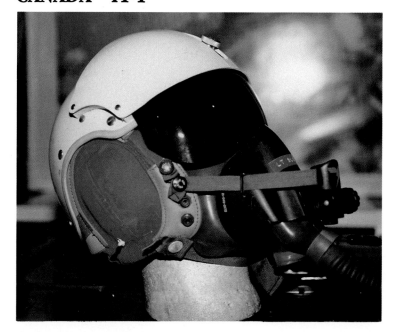

The 41-1 helmet was first issued in the early 1970s for use in several types of aircraft. The Gentex helmet offers a unique close-fitting design very similar to an armor/tankers helmet. The fiberglass 41-1 helmet has a reflective cross on top and the option of a dark or clear visor. The MS22001 oxygen mask snaps on and utilizes a unique ratchet-strap tension system.

CANADA - SPH-4 (CF)

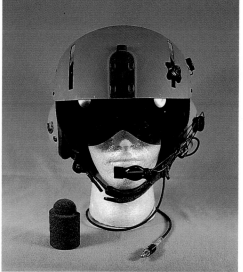
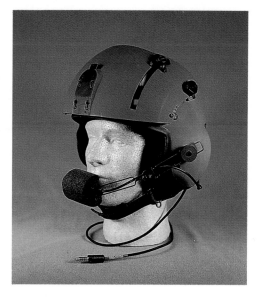

Produced in the U.S. by Gentex, the SPH-4 is the newest issue helicopter helmet for Canadian forces. The rigid visor housing permits attachment of night vision goggles. Thick foam microphone cover filters ambient noise.

CHINA - TYPE 25

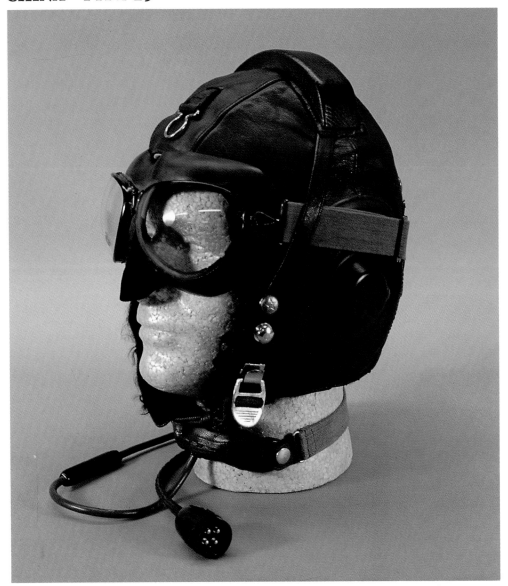

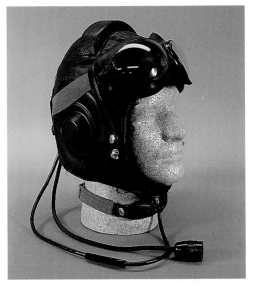

The 1950s Type 25 leather helmet is very similar to the Russian ShZ-50 leather helmet and was issued in both winter fur-lined and summer silk-lined models. The helmet was used with throat microphones and goggles. Goggles are Russian design and have dark lens option. The type 25 was used extensively during the Korean Conflict and by North Vietnamese pilots flying MiG-17 fighters. When used in fighters such as the MiG-15, an oxygen mask directly copied from the Russian KM-15M is worn. The KM-15M is a copy of the German 10-67 mask. Electrical connector is a copy of the Russian model for compatibility with Russian aircraft. The Russians copied it from the Germans.

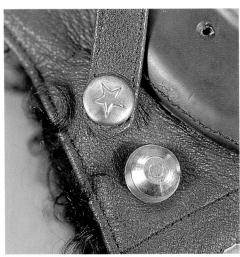

LEFT: Snaps have distinctive star.

CHINA - TK-1 (1967 MODEL)

The TK-1 is a copy of the high-altitude Russian GSh-4AA, which was most likely copied in basic design from the U.S. K-1/MA-2. Used by North Vietnamese pilots flying MiG-21 fighters, the TK-1 saw combat during the Vietnam Conflict.

Oxygen first enters around interior of visor to prevent fogging.

Communications and oxygen connections are compatible with Russian aircraft.

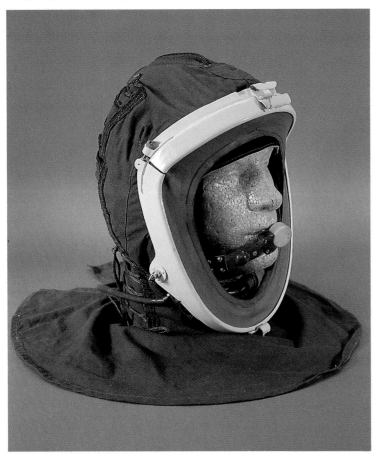
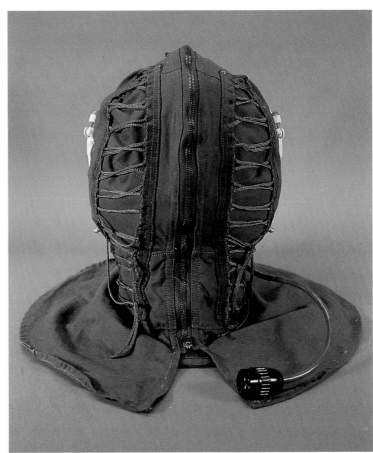

The microphone is adjustable in length. Neck skirt is tucked under DC-1 or DC-2 partial-pressure suit. The TK-1 is pressure sealed by way of a thin surgical rubber hood supported by a clothe hood which is adjustable by laces on each side of the center zipper.

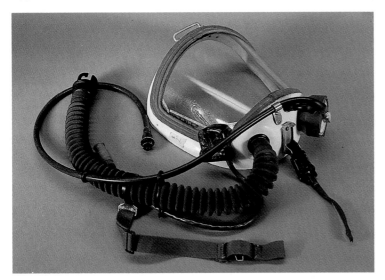

ABOVE: The TK-1 uses an electrically heated face plate to prevent fogging. The small black tube is the emergency oxygen line and feeds directly into a regulator mounted on the removable face plate.

RIGHT: All white segments of helmet are aluminum. Quality of paint is poor.

CHAPTER 3: FOREIGN FLIGHT HELMETS - CHINA

Issued wooden container is similar to Russian.

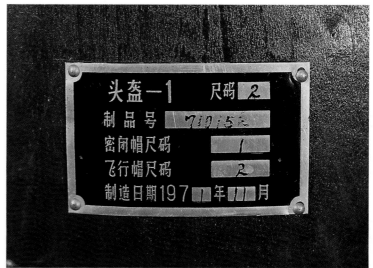

Container information plaque.

Identification on neck skirt indicates this example was produced in 1971.

CHINA - TK-1 (1985 MODEL)

Improved version of earlier TK-1. The 1985 Model TK-1 is worn with the DC-3 partial-pressure suit.

White plastic parts replace more fragile bakalite.

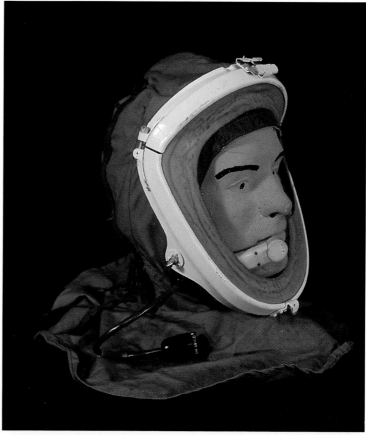
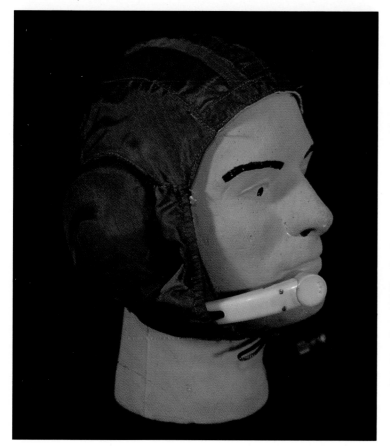
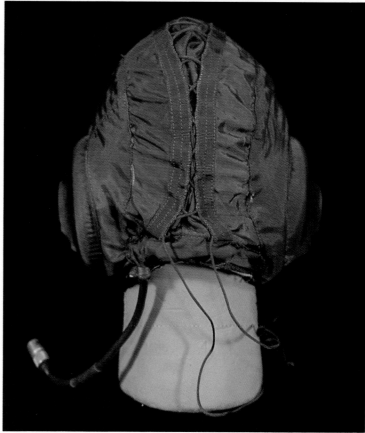

Inner communications helmet remains the same but bakalite boom microphone is replaced by trimmer, plastic model.

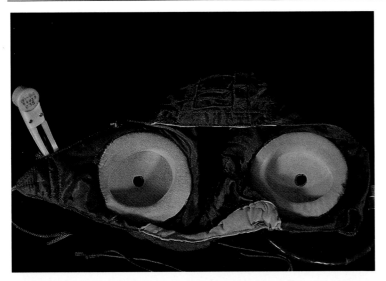

Construction remains aluminum.

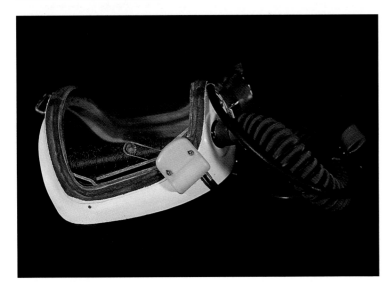

CHAPTER 3: FOREIGN FLIGHT HELMETS - CHINA

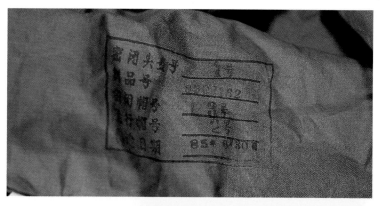

BELOW:
The TK-1 outfit with container illustrates protective shipping bags and manual.

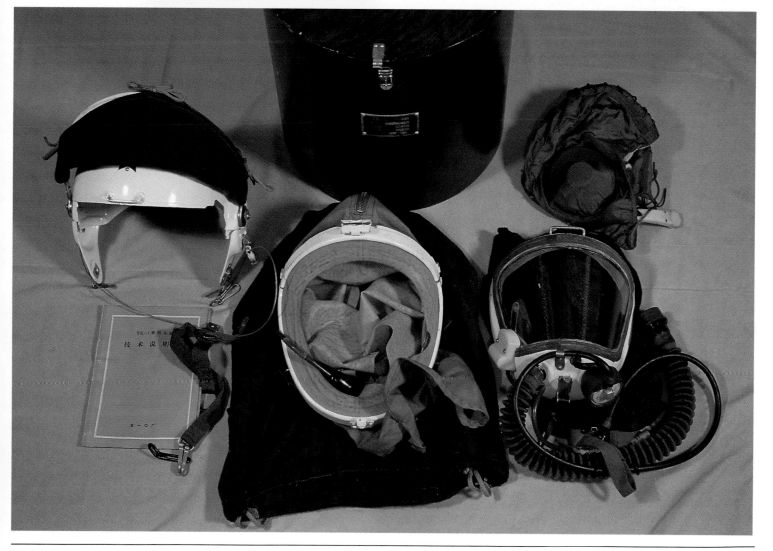

CHINA - TK-2

1970s fiberglass fighter helmet has single, dark visor. Holes in visor housing reduce dangerous wind drag during ejection. Used in all Chinese fighter aircraft, the TK-2 has connections compatible with Russian aircraft. Oxygen mask is similar to Russian KM-32.

CHAPTER 3: FOREIGN FLIGHT HELMETS - CHINA

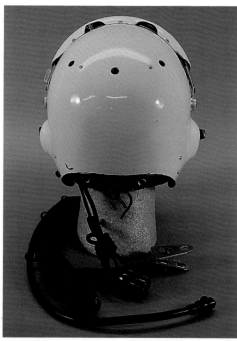
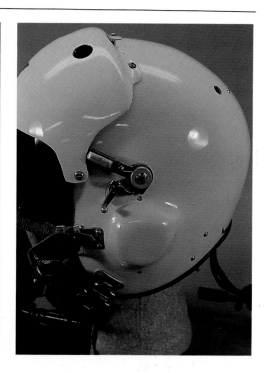

MiG-21 pilot.

FRANCE - TYPE 403 M3

Designed from the U.S. H4, this army helicopter model was produced by Socapex in France. Used with a boom microphone in the 1950s by Alouette crews.

FRANCE - GUEREAU 312

Used in the 1960s in fighters. Cloth interior, holes provide ventilation and reduce drag upon ejection. Fiberglass helmet uses Ulmer MS22001 type oxygen mask. This example was used by the 2/12 "Cornovailles" Intercepter Squadron flying Dassault Super Mystere B2.

FRANCE - GUEREAU 313M1

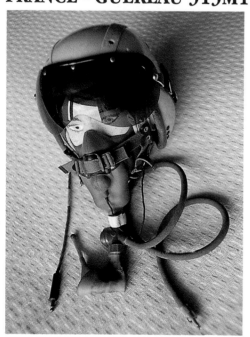
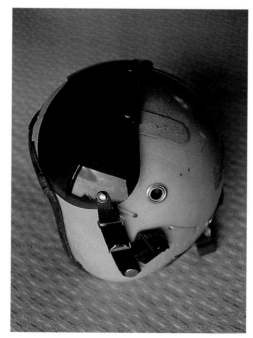
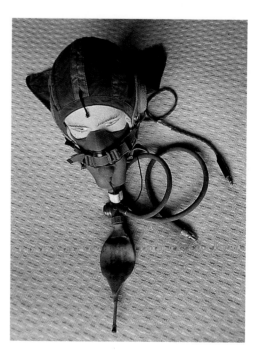

Used by the French Navy beginning in the late 1950s. Oxygen mask is Ulmer MS22001 type. Inner communications helmet is green clothe and leather.

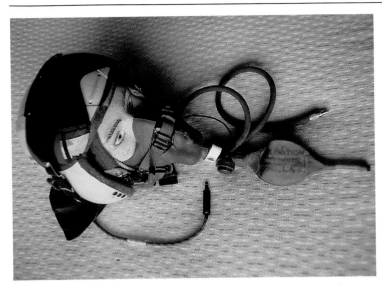

Fiberglass helmet has interior laces for universal fitting.

FRANCE - GUEREAU 316

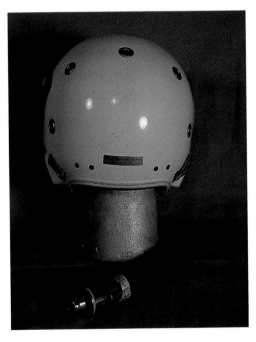

1980 Model has side-actuated dual visors. Clear visor is spring-loaded and is either full up or down. Dark visor is adjustable, but use require clear to be down.

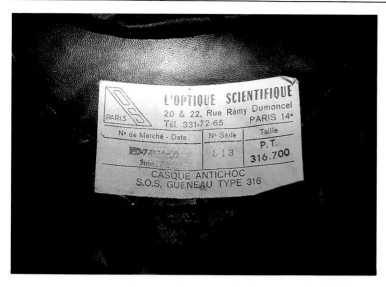

Used by the 3/5 "Combat Venessain" Intercepter Squadron flying Dassault Mirage F-1.

LEFT AND ABOVE: 1985 Model with Ulmer type 82M oxygen mask. Used by 4/11 Fighter Squadron flying Jaguar A fighters.

Ulmer oxygen mask type 82 doubles as chin strap.

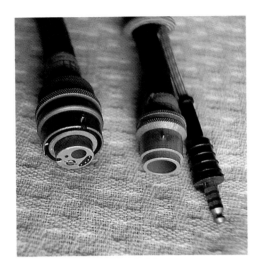

Type 316 with French manufactured SOPELEM Model CN2H Night Vision Goggles as used by the French Army and Air Force by helicopter crews.

1980s Helicopter version has boom microphone and amber visor replaces the clear.

FRANCE - GUEREAU TYPE 478

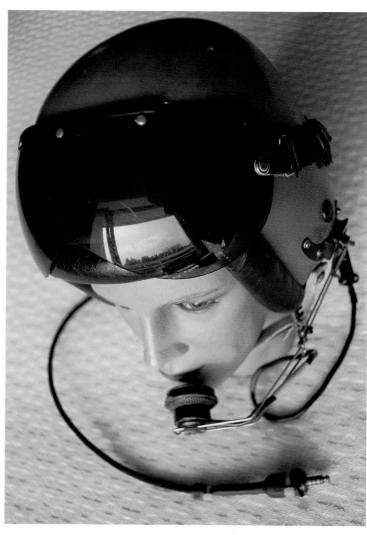

Used by the French Navy beginning in 1978 for search and rescue Super Frelon helicopter crews. Dark visor is held on by snaps and straps, shell is fiberglass.

FRANCE - TYPE 12

Developed to replace the K-1 used by the French Air Force in the 1950s and roughly based on the U.S. helmet, the high-altitude Type 12 differs primarily by having the oxygen hose connect to the helmet frame as opposed to the removable face plate. Function, fit, and structural design are similar to the K-1. Produced by EFA in 1958 for use in Mirage III, IV, and V aircraft it has a fiberglass shell. Reportedly, production was limited to under 450. Neck skirt tucks into EFA-ARZ Type 30 partial-pressure suit.

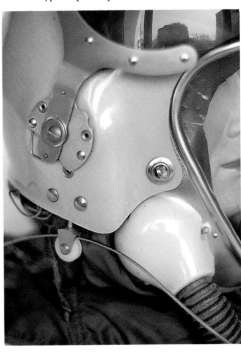

With protective fiberglass shell removed.

Closeup of oxygen intake, hold down cable system, and tinted visor mount.

CHAPTER 3: FOREIGN FLIGHT HELMETS - FRANCE

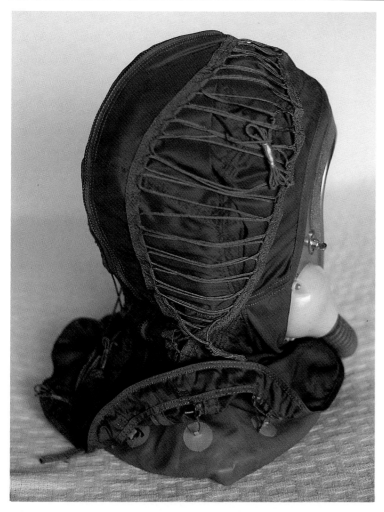 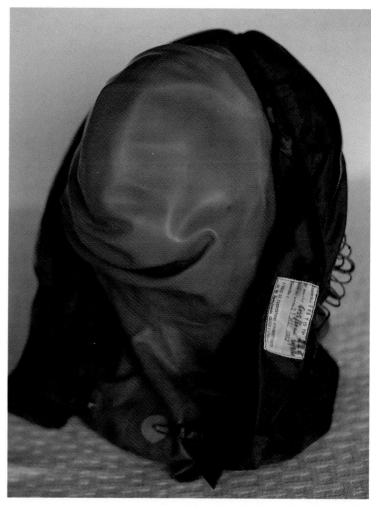

Nylon hood supports surgical rubber bladder and is adjustable by way of laces on both sides. Pressure containment hood is similar to K-1, MA-2, Soviet GSh-4, and Chinese TK-1.

Communications connection and hold down cables. Unlike the K-1 or MA-2, the Type 12 plexiglass face plate has no defogging feature or oxygen intake. EFA decal.

FRANCE - TYPE 21/23

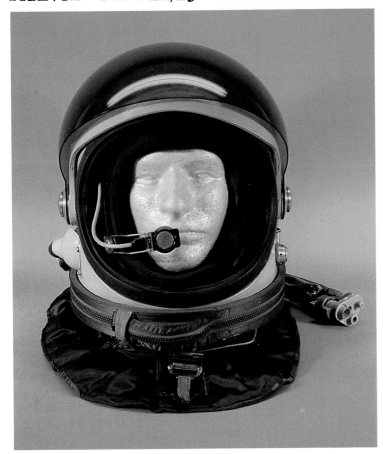

First used in 1967, the fiberglass Type 21 was produced by EFA for high-altitude with the EFA-ARZ Type 30 partial-pressure suit.

Left pivot button controls locking of main visor in either full up or down positions, right pivot button controls dark visor in the same manner. Clear visor is double-walled with a 24-volt defogging system.

CHAPTER 3: FOREIGN FLIGHT HELMETS - FRANCE

The Type 21 was also used by the air forces of Germany, Morocco, and Switzerland.

The Type 21 modified for use by the Belgian and Dutch Air Forces for the F-104 Starfighter are designated the Type 23. Primarily the oxygen system is changed.

Visor is sealed by inflatable gasket system activated by valve when clear visor is lowered. To deactivate, stainless steel button is depressed.

Neck ring is attached to helmet by zipper.

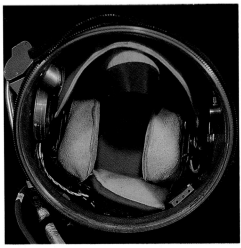
The Type 21 is fitted by tension-adjusted nylon cord system similar to U.S. HGK-19. Pads are very soft leather, face seal is laminated rubber similar to diver's wet suit material.

Stainless steel fittings and hardware abound including the regulator, seen at the bottom.

Label on neck ring.

Oxygen connector on Type 21.

Oxygen connector on Type 23.

Issued carry case is heavily padded.

GERMANY - GUEREAU TYPE 316

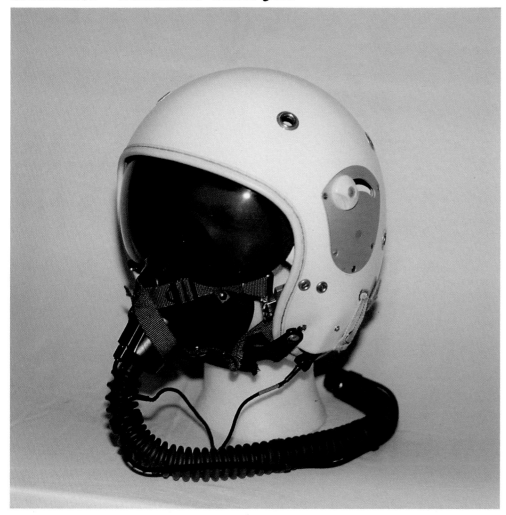
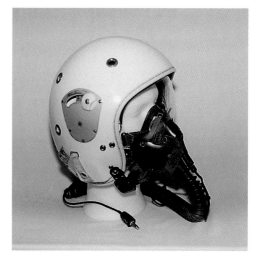

French-produced fiberglass helmet uses German-produced MBU-5/P oxygen mask. Used by fighter pilots in the 1970s-1980s, the Type 316 has dual visors. Unlike French version, German model uses chin strap.

GERMANY - HGU-55/G

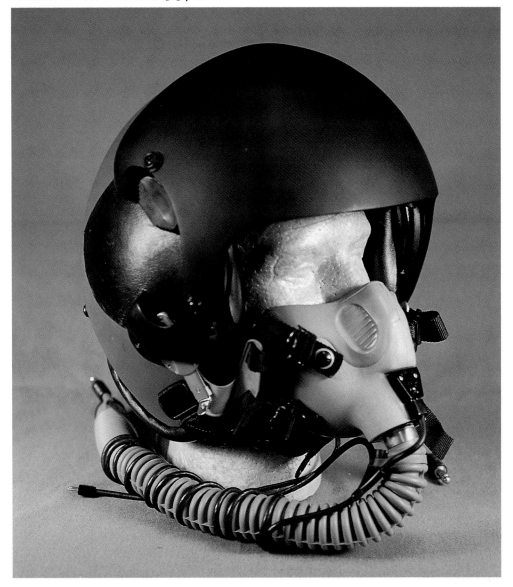

Adopted by the Luftwaffe in 1980, Dual Visor Model produced in the U.S. by Gentex, uses MBU-12/p oxygen mask. Visor housing extends to bayonet receiver to prevent snagging of parachute straps during ejection. Used in F-4 Phantom II fighters.

Improved 1990 Dual Visor Model uses MBU-12 mask with "WEA" water-activated bayonet on one side. System automatically releases mask in water to prevent suffocation by injured pilot after ejection over water. Visors are both activated from the left side so pilot can always keep right hand on control stick.

POLAND

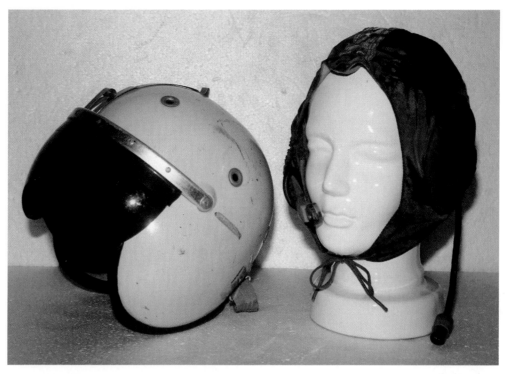

Used in the 1960s, this fiberglass helicopter helmet uses the Russian GSh-4 inner communications helmet with boom microphone.

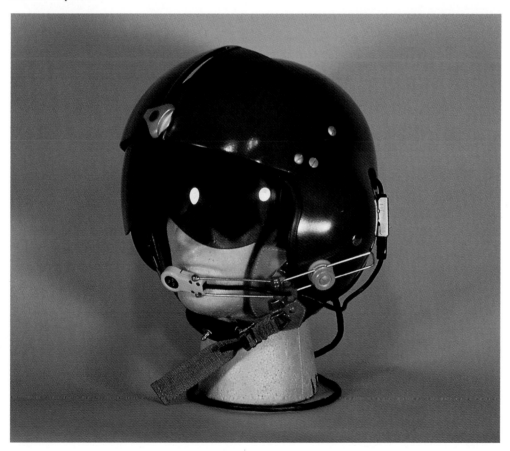

Helicopter helmet first issued in early 1980s. Very similar design to U.S. SPH-3, uses Russian boom microphone or Russian LA-3 throat microphone, and Russian KM-32 oxygen mask. Fiberglass shell, used in Mi-8 and Mi-24 helicopters.

RUSSIA/SOVIET UNION - ShZ-50 SERIES

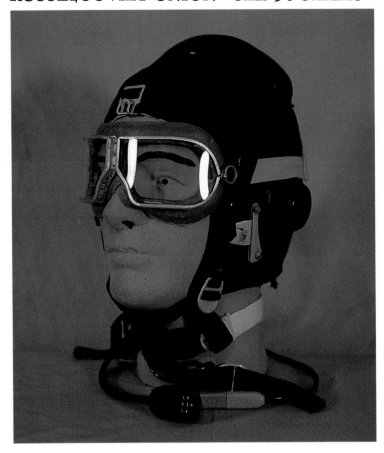
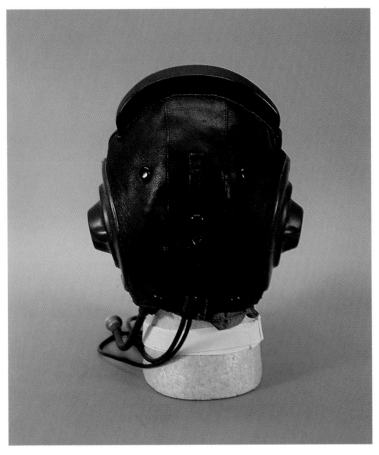

Modified from World War II models and introduced in 1950, the ShZ-50 is fur-lined winter-issue leather. These helmets were worn in all types of aircraft including MiG-15 fighters and helicopters. They are still used for sport and student flying plus helicopters. The ShL-50 is the same but for a cloth lining for warmer climates. All models use the same design communications plug as German World War II helmets. BELOW: A mesh version of the ShL-50 is for hot climates. All versions can use the KM-15 mask, which is a copy of the World War II era German 10-67 mask.

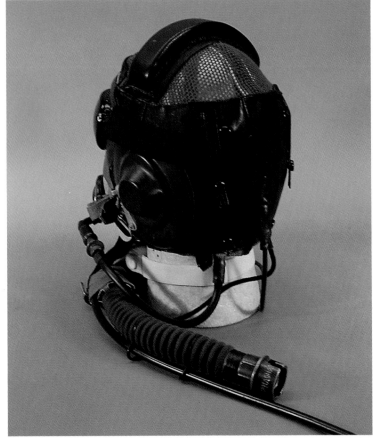

CHAPTER 3: FOREIGN FLIGHT HELMETS - RUSSIA/SOVIET UNION

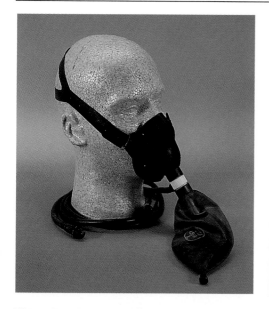 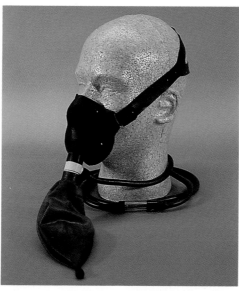 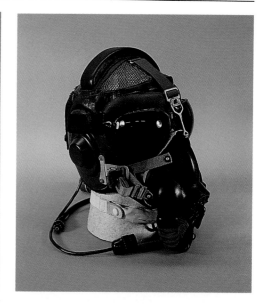

TOP LEFT AND CENTER: The KM-15 mask is held in place by a single rubber strap. TOP RIGHT: Newer ShL-61, ShL-82, ShZ-61 and ShZ-82 helmets are modified with additional straps for use with the ZSh-3 helmet but are still in use alone, often mated with a KM-32 oxygen mask which is produced in black and green.

ABOVE: Both types of goggles are World War II models and are still in use today, often worn by students.

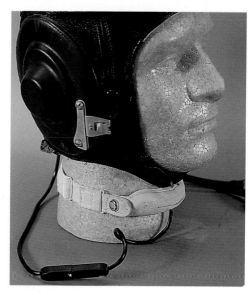

LA-3 throat microphones are used with many Russian helmets.

LEFT: ShL-50 and KM-15 worn by Chinese MiG-15 pilot during Korean Conflict.

157

RUSSIA/SOVIET UNION - ZSh-3 SERIES

First issued in the early 1960s, the basis for the ZSh-3 is formed by a ShZ-50 leather helmet with additional straps added to hold a rigid, aluminum shell in place. The shell is padded and has a notch in the top to mate with the single padded ridge of the leather helmet to hold both in place.

The large visor has three positions: Up, Down, and Middle. The three holes in the crown are to reduce wind drag upon ejection.

A KM-32 oxygen mask is held in place by way of adjustable straps on the sides, a metal bracket on top, and a press stud on the bottom. Some KM-32 masks have an integrated microphone, but use of throat microphones are more common.

CHAPTER 3: FOREIGN FLIGHT HELMETS - RUSSIA/SOVIET UNION

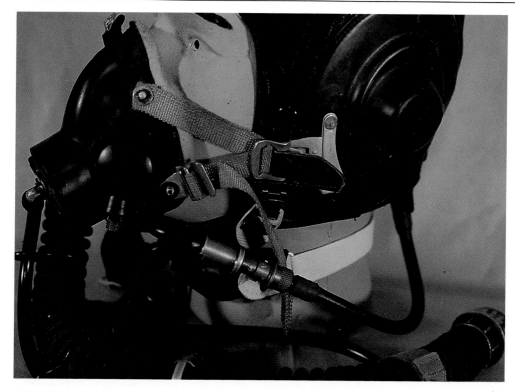

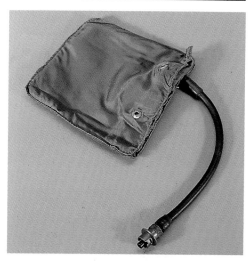

The air bag consists of a rubber bladder protected by a nylon bag and held in place with snaps.

ABOVE: The ZSh-3M uses an air bag which fits against the back of the head and is pressurized through the oxygen mask system. This system tightens the face against the mask automatically as the oxygen pressure is increased. This system was developed by the Russians and is used on all of their fighter helmets. It has been adopted by Sweden, and is now being used by the U.S. in the "Combat Edge" system.

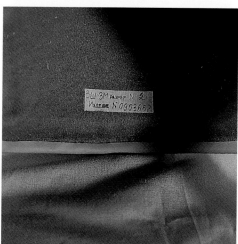

The most widely used helmet in the world, the ZSh-3 is worn here by Bulgarian pilots.

East German pilots were known for their orange painted ZSh-3's. The bright color aided in rescue from the freezing Baltic Sea — the former East German Air Force tactical training area.

MiG-23M pilot wearing an early ZSh-3 with the unpopular aqua blue visor. Worn in virtually every type of Russian aircraft from MiG-17 to MiG-29 fighters, to Il-76 cargo planes to Mi-24 attack helicopters, the ZSh-3 is also exported for all Soviet aircraft and therefore worn in more countries than any other jet-age helmet.

In the combat of Afghanistan, a special ZSh-3 was issued to attack helicopter crews. It was constructed of ballistic titanium and mounted a shatter proof clear visor.

The armor helmet was worn with mesh ShL-82 helmet.

CHAPTER 3: FOREIGN FLIGHT HELMETS - RUSSIA/SOVIET UNION

Still in production in 1990, the latest ZSh-3M PS model is issued in typical Russian wooden container with KM-32 mask, ShZ-82 Winter helmet, ShL-82 Summer helmet, LA-5 throat microphone, and clear lens goggles for night flying.

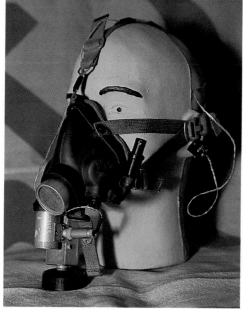

ABOVE:
Unusual KM-32 mask modified with tiny turbine and electronics to measure pilot's breathing capacity and rate.

LEFT: Issued wooden shipping container.

161

RUSSIA/SOVIET UNION - ZSh-5 SERIES

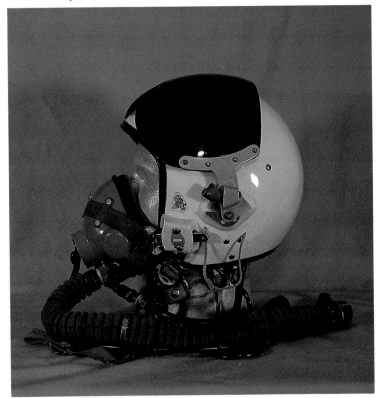

The one-piece, fiberglass ZSh-5 helmets were first issued in the early 1970s for use in combat aircraft.

The fiberglass shell KM-34 oxygen mask offers improved breathing over the smaller KM-32. Mask has microphone mount but LA-3 throat microphones are more popular.

Some early models used ineffective aqua blue "Arctic" visor which was soon replaced by green and later brown tints.

CHAPTER 3: FOREIGN FLIGHT HELMETS - RUSSIA/SOVIET UNION

The ZSh-5M began service in the late 1970s and is used with the VKK-3M flight suit. The main improvement was the addition of an electronic visor actuation system.

Visor actuation system instantly lowers the visor upon ejection. Wired into the ejection system, this permits night or poor weather flying without the need for a clear visor which protects the pilot from the severe wind blasts experienced during ejection.

RIGHT: Unusual and rare helicopter model ZSh-5 with dual visors and boom microphone from GSh-6A high-altitude helmet.

The ZSh-5MKV-2 has mounts on top for the Sh-3UM-1 target designator system found on MiG-29, MiG-33, Su-27, and Su-30 fighters.

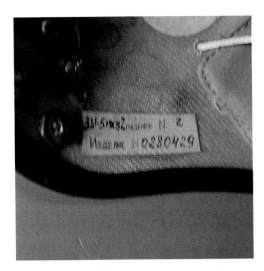

MiG-23 pilots wearing ZSh-3's (left) and ZSh-5's (right), three are early models with "Arctic" blue visors Helmet types are often mixed within the same squadron, even the same aircraft.

LEFT: ZSh-5's with frontal aviation MiG-25 pilots.

The ZSh-5A PS is 1992 issued and uses the KM-34D Series 2 "Side Hose" mask. It is used with the VKK-6D, VKK-15, or BMCK-4 flight suits and uses an LA-5 throat microphone, or mask mounted microphone.

Issue cold weather insulation hood.

LEFT: The ZSh-5A PS is issued with mask and storage bag, cold weather hood, cotton skull cap, LA-5 throat mike, and spare chin strap, brow, ear, and top pads.

RUSSIA/SOVIET UNION - ZSh-7 SERIES

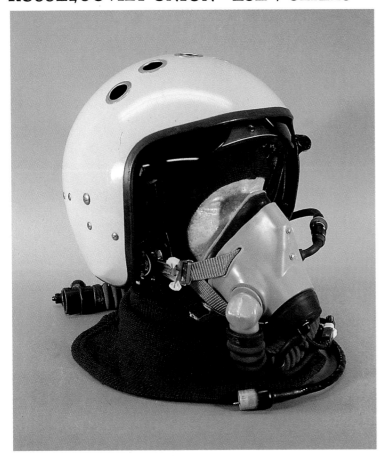 

First issued in 1987, the ZSh-7A initially used the KM-34 mask until the KM-34D Series 2 shown here became available in 1989. ZSh-7A uses air bag system similar to ZSh-5. Holes relieve drag experienced during ejection. LA-5 throat microphones remain popular, though the KM-34D Series 2 mask with microphone is now standard. A single dark visor is used along with an automatic visor deployment system pioneered on the ZSh-5M.

CHAPTER 3: FOREIGN FLIGHT HELMETS - RUSSIA/SOVIET UNION

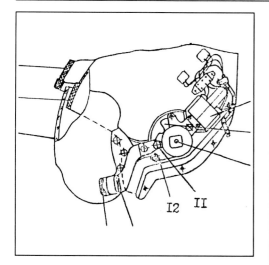

Diagram of the automatic visor lowering device which is linked to the ejection system.

Mounting bracket for target designator is adjustable in the X and Y axis to attain proper alignment with the pilot's eye.

The Sh-3UM-1 target designator is a helmet-mounted aiming system for use in air to air combat.

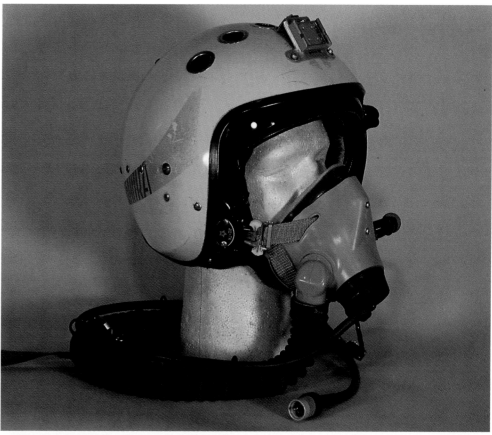

ABOVE AND BELOW: The ZSh-7LP was introduced in 1990. It is designed for use with the NVE or Sh-3UM-1 target designator system and uses the KM-34D Series 2 mask.

167

JET AGE FLIGHT HELMETS

The ZSh-7APN was introduced in 1992. It is also designed for use with the NVE or Sh-3UM-1 target designator system but uses the newer KM-35 mask with integrated microphone for use in MiG-29M, MiG-33, Su-30MK, Su-32FN, and Su-35 aircraft.

MiG-29 pilots wearing ZSh-7A. VKK-15 flight suits are generally worn under a variety of two-piece outer suits in black (as seen here), navy, grey, tan, or camouflage. For extreme conditions, the orange BMCK-4 exposure suit system is used.

ABOVE: Connector of KM-35 is different than other Russian models and not interchangeable. BELOW: Label of KM-35 is located inside plastic shell.

KM-35 is lightweight and has microphone.

CHAPTER 3: FOREIGN FLIGHT HELMETS - RUSSIA/SOVIET UNION

The ZSh-7LP is issued with mask and storage bag, cold weather hood, cotton skull cap, LA-5 throat mike, and spare chin strap, brow, ear, and top pads.

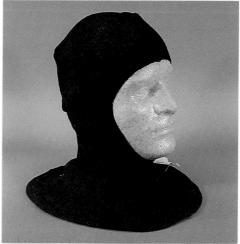

A hood is issued for winter flying.

RUSSIA/SOVIET UNION - GSh-4 SERIES

The GSh-4 was introduced in the late 1950s and appears to owe much of its design to the U.S. K-1/MA-2 It differs in the mounting of the protective outer shell, which is aluminum. The frame and the face plate/visor frame are also aluminum with thin, surgical rubber serving as the pressure seal and seal for the face plate. Shown here without the face plate, the GSh-4 was used for high-altitude flight in MiG-19, MiG-21, and Su-15 fighters. The GSh-4M model was introduced in 1960. The primary improvement over the GSh-4 is the addition of the dark visor. RIGHT: Visor is electrically heated and, also like the MA-2, hooks at the top and latches at the bottom.

CHAPTER 3: FOREIGN FLIGHT HELMETS - RUSSIA/SOVIET UNION

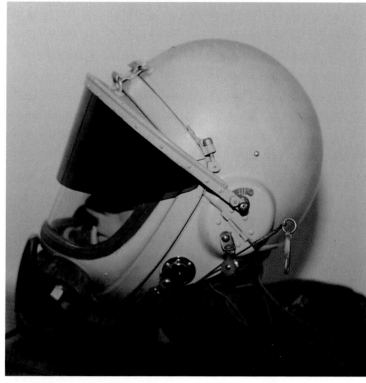

Dark visor can be positioned as needed.

The neck skirt tucks into a VKK-4 partial-pressure suit.

Oxygen first enters around interior of visor through rubber ducting to prevent visor fogging.

The inner communications helmet has adjustment laces on both sides.

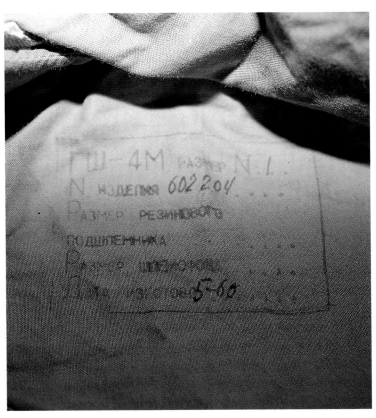

Label on neck skirt indicates May, 1960 manufacture date, dispelling the belief helmet type was copied from Francis Gary Power's particular MA-2 following his May 1960 U-2 shoot down over the Soviet Union.

GSh-4AA model was introduced in 1961. Primary improvement over the GSh-4M are the reinforcement ribs formed in the aluminum shell to add rigidity and thus improve protection. Visor heating electrics are also improved. The small black tube on the bottom of the face plate is an emergency oxygen tube which feeds directly into the regulator.

RUSSIA/SOVIET UNION - GSh-6 SERIES

The high-altitude GSh-6M was introduced in 1962 to replace the GSh-4 series. Emergency oxygen regulator is mounted on left side below visor pivot. The GSh-6M was used in Su-15, MiG-21, and MiG-23 aircraft. Shell is aluminum and heavily padded for protection, comfort, and insulation. Visor is electrically heated. Oxygen enters helmet first around visor to prevent fogging.

  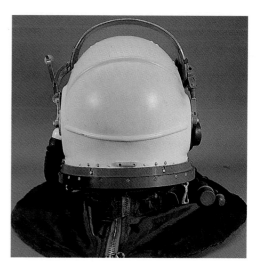

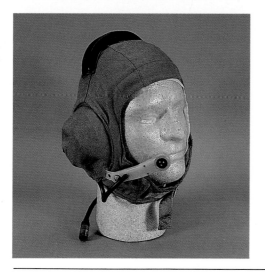 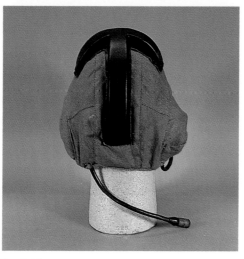

TWO AT LEFT: Inner communications helmet in made of fire-resistant material. Padded ear cups are leather. The black leather pads fit into designed notches of helmet for proper fit.

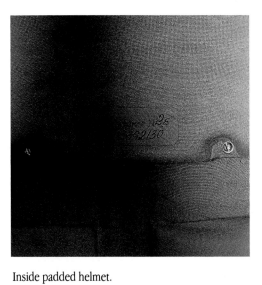

A lever controls the position of the internal dark visor.

Inside padded helmet.

ABOVE LEFT: Aluminum neck ring latches to helmet with six lugs and seals helmet around neck by use of surgical rubber collar. Skirt tucks into VKK-4M or VKK-6M partial-pressure suit. CENTER: Visor is mechanically sealed by cam action of locking lever.

Czech MiG-23M pilots wear GSh-6M (left) and GSh-6A (right).

Introduced in 1968, the GSh-6A differs from GSh-6M in several ways. Most notable is the oxygen intake and emergency oxygen regulator are reversed. It is used in Su-15, MiG-21, MiG-23 and MiG-25 fighters. The two-part rotating neck ring is replaced by one-piece. Orange skirt tucks into BMCK-4/VKK-6MP-1 partial-pressure suit. GSh-6AA, introduced in 1980, differs only in that it has 8 pins instead of 6 connector pins to attach to neck ring. It is used in the MiG-31.

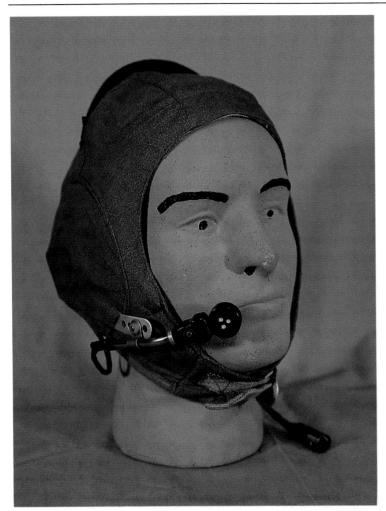
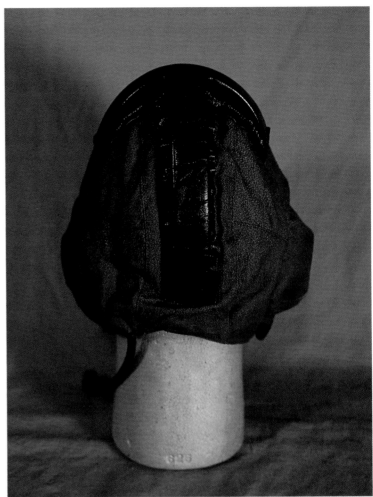

Boom microphone is adjustable.

MiG-25 pilots wearing GSh-6A helmets and VKK-6M partial-pressure suits.

SWEDEN - H4

LEFT: Swedish Air Force used re-wired U.S. produced H4 with MS22001 mask in their first Saab jet aircraft. RIGHT: Modified with mounted dark visor.

SWEDEN - MODEL 106

 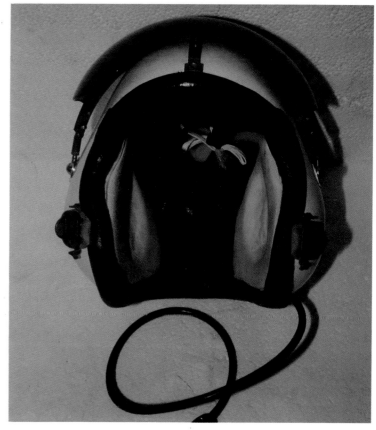

In the mid-1950s, FFV Aerotech AB began producing the model 106 in Sweden. It was nicknamed "Bulan" and was issued with only a dark visor. It used an M7349-111000 mask.

SWEDEN - K-1/MA-2

Sweden filled the need for high-altitude headgear in the 1950s by utilizing re-worked U.S. K-1 helmets. Oxygen intake and wiring are modified.

U.S. produced MA-2, reworked in the same manner as the K-1, but with the additional modification of a Swedish microphone was used in Drakken fighters.

SWEDEN - MODEL 111

Single visor helmet produced in three sizes: 1, 2, 3. With permanent liner and uses M7349-111000 oxygen mask.

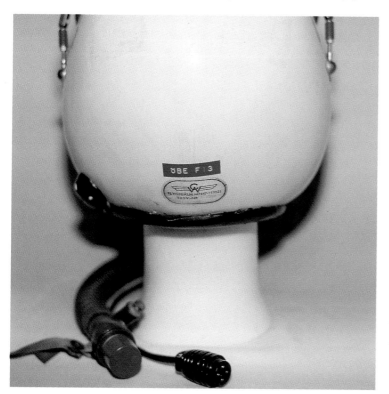

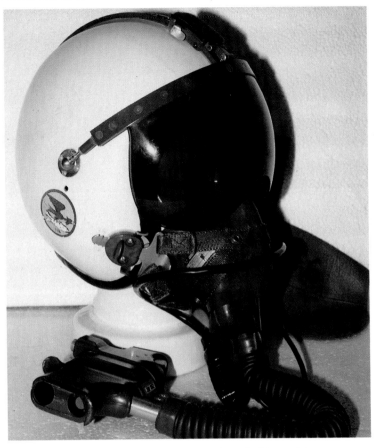

Swedish oxygen systems place the regulator on the end of the mask hose.

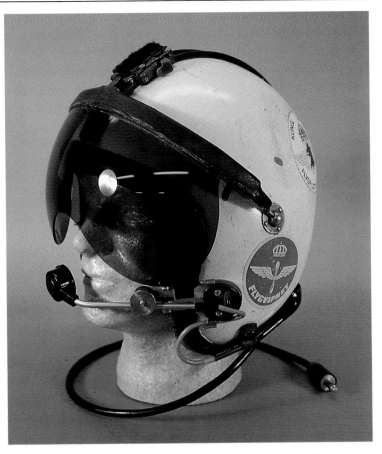

ABOVE AND BELOW: Helicopter model uses boom microphone and chin strap, both attached by way of mask receivers.

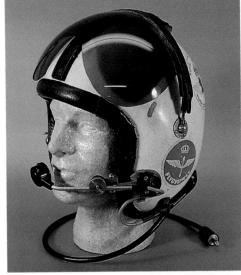
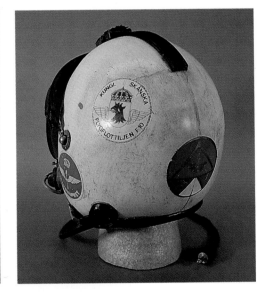

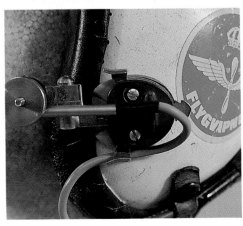
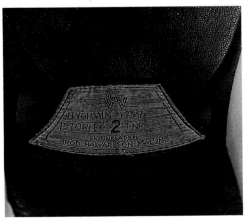

CHAPTER 3: FOREIGN FLIGHT HELMETS - SWEDEN

SWEDEN - MODEL 112B

Used in fighter aircraft, the 112B has an air bag system similar to the Russian ZSh-5M. Dual visors are optional.

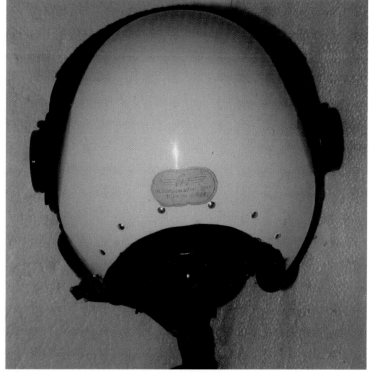

Oxygen mask is model M7349-112000. Tube leading from mask to air bag can be seen below bayonet fitting.

SWEDEN - MODEL 113

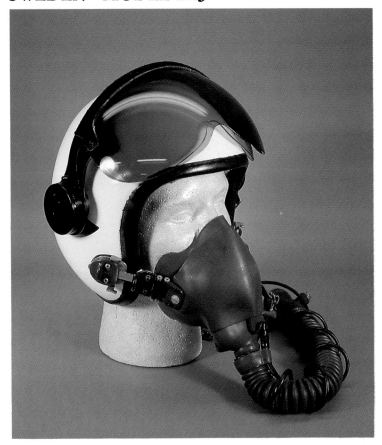

Optional dual visor Model 113A with Type MS22001 Mask. Model 113A is issued in three sizes and used in all Swedish aircraft in the 1970s.

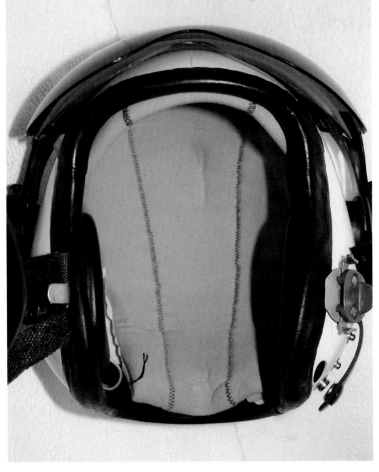

Padded cloth interior is custom fit.

CHAPTER 3: FOREIGN FLIGHT HELMETS - SWEDEN

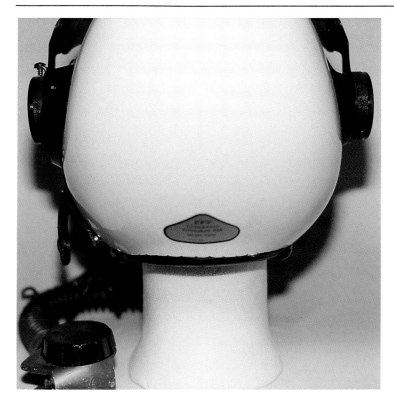

The Model 113B is similar to Model 113A, but with air bag system.

Dual visors on Model 113B are fully independent in operation.

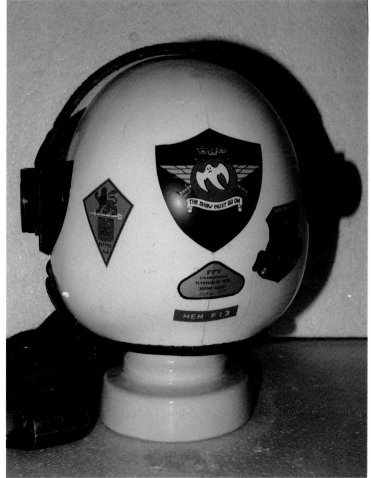

SWEDEN - MODEL 114

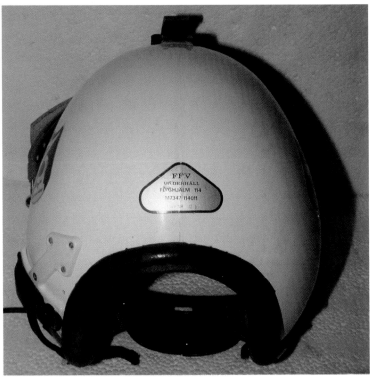

Used in helicopters, the Model 114 is basically a Model 113 with a Model 111 single visor. As with all Swedish-produced helmets, the Model 114 is made by FFV and is issued in three sizes.

SWEDEN - MODEL 115

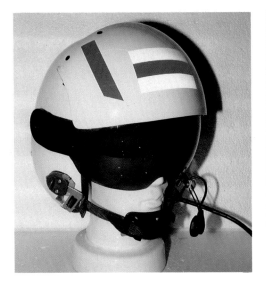
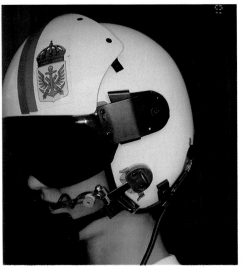
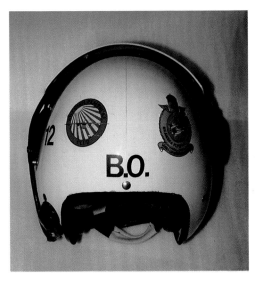

1980s issue helicopter model issued two sizes: 2, 3. Single dark visor is protected by housing. Boom microphone is mounted by way of oxygen mask receiver.

SWEDEN - MODEL 116

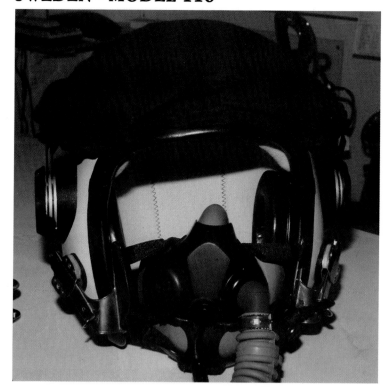
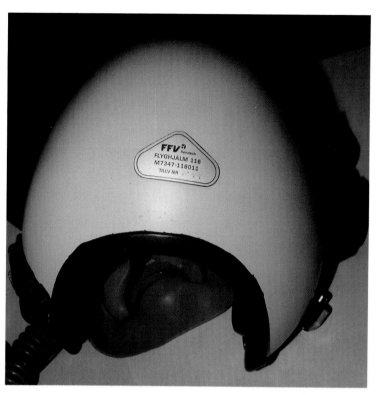

Current issue lightweight helmet for use in fighters. Uses dual visors and U.S. produced Gentex MBU-20/P oxygen mask.

Chemical/biological protective model uses U.S. Gentex HA/LP-PPB oxygen mask. Has drinking capability and can be used in fighters, helicopters, and transports.

SWEDEN - MODEL 117

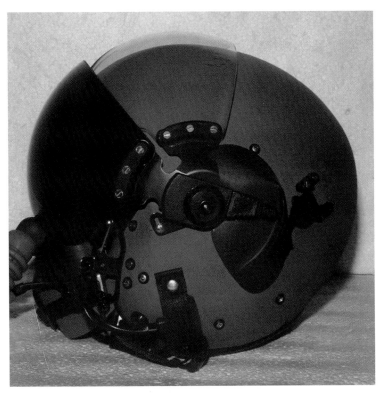

UK produced Helmets Ltd. "ALPHA." In service but being replaced by Model 116. Dual visors are independent.

Mask is U.S. MBU-20/P.

Boom microphone is an option.

SWEDEN - MODEL 119A

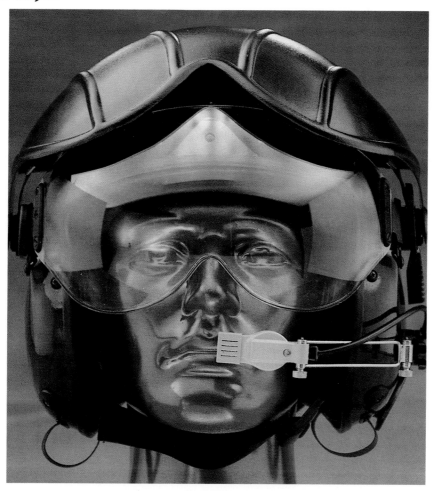

UK produced Helmets Ltd. "Alpha." Helicopter model used by Swedish Army.

UNITED KINGDOM - MK.IA/M.

Produced by Helmets Ltd., the fiberglass MK.IA/M. entered RAF service in the early 1960s for use in most combat aircraft. It is worn over an inner helmet and has a single dark visor which is mounted solely on the single center track. Visors were available in light or dark tints. The "H" Type oxygen mask was produced in green and black and issued in three sizes.

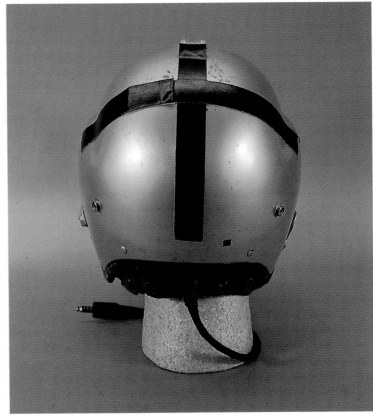

CHAPTER 3: FOREIGN FLIGHT HELMETS - UNITED KINGDOM

Helmet was offered in five shell sizes: 1, 1.5, 2, 3, 4.

Inner "G" helmet was offered in four sizes and two colors: Blue/grey and Olive/drab.

Green model with reflective cross was used into the 1970s by RAF Lightning pilots of 5 Squadron.

UNITED KINGDOM - MK.IIA

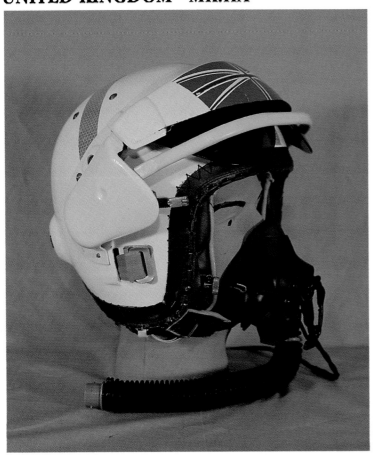

Entering service with the Royal Navy beginning in 1967, the fiberglass MK.IIA was used with a black "P" type mask. Single visor was available in light or dark tints. "P" mask is adjusted by way of threaded knobs on both sides which offer infinite tension settings. Later versions mounted a protective visor cover.

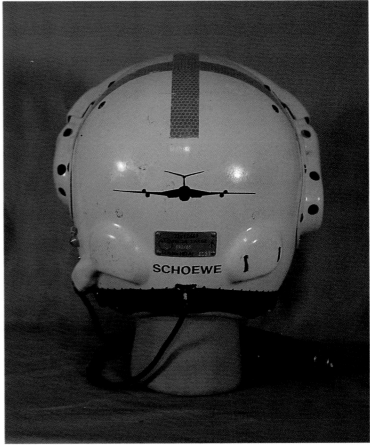

CHAPTER 3: FOREIGN FLIGHT HELMETS - UNITED KINGDOM

TOP LEFT: Helmets Ltd. developed the intricate energy-absorbing liner which is used on many later model helmets through the "Alpha" series.

RIGHT: Laced on edge rolling is complex, as is liner, but has proven effective and durable.

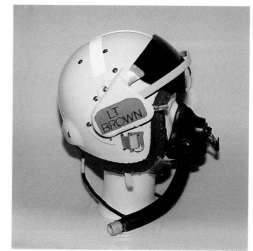

Early model without protective visor cover. Side buckles tension interior lacing for a snug fit.

UNITED KINGDOM - MK.IIIA/B

Adopted by the RAF in 1967, the MK.IIIB is basically a Helmets Ltd. MK.IIA with a MK.I visor assembly and fitted with a "P" mask. It will also accept "Q" and "H" masks as well.

Side buckles tension interior lacing for a snug fit.

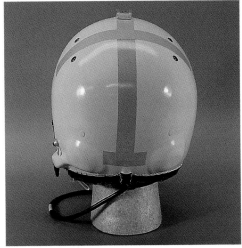

Black "P" mask is mounted with the same threaded tension system as MK.IIA.

CHAPTER 3: FOREIGN FLIGHT HELMETS - UNITED KINGDOM

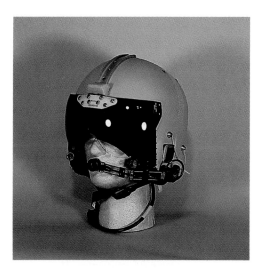

For use in RAF and Royal Army helicopters, the MK.IIIA was offered with a boom microphone.

First issued in the 1970s, this Helmets Ltd. fighter helmet remains popular. Used in all RAF combat aircraft, the MK.IIIC mounts an independent dual visor and a black/green "P" mask. Protective clothe visor cover is often left in place during flight. Edge rolling can be black, green, or tan.

The fiberglass shell is offered in green or desert tan.

UNITED KINGDOM - MK.IV

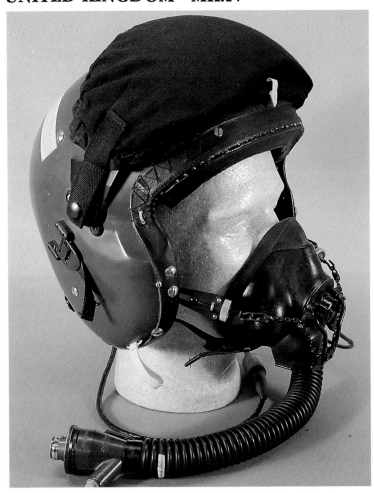
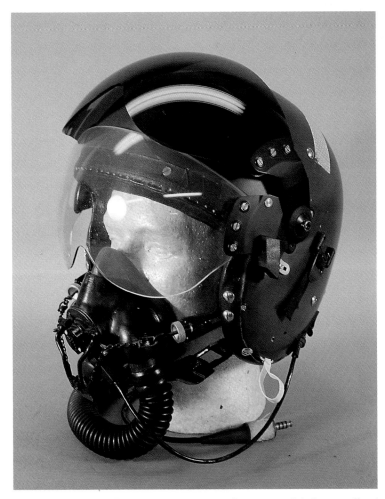

The MK.IV was intended to fully replace all remaining MK.I, MK.II and MK.III helmets in British service for both fixed and rotary wing aircraft. To accomplish this, it is offered in many variations. All attachments are held on by screws to facilitate interchangability and upgrades. RIGHT: Visors operate independently. BELOW RIGHT: White strap extending from bottom of helmet adjusts earphones for comfort once helmet is in place.

CHAPTER 3: FOREIGN FLIGHT HELMETS - UNITED KINGDOM

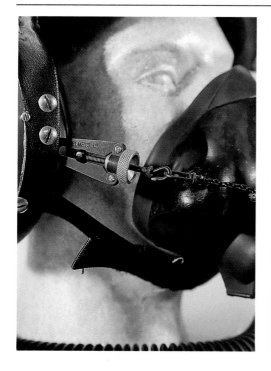

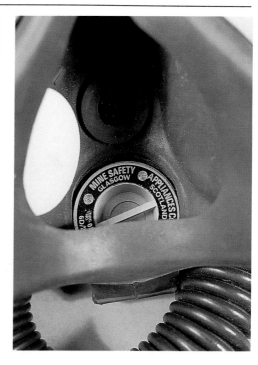

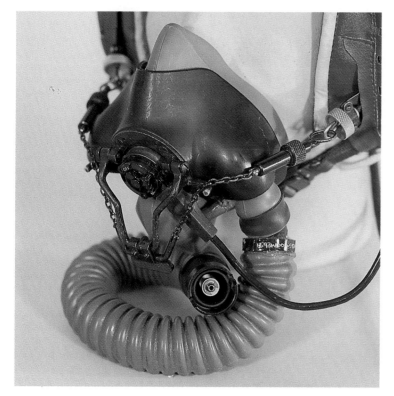
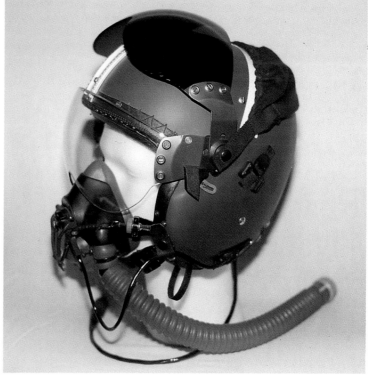

1980 RAF MK-IVA with "P" mask. Visors move further back on helmet to allow for NVG mount.

ABOVE AND BELOW: RAF MK-IVB with "P" mask as used by Red Arrows in 1987. Visors are side operated.

RIGHT: The standard RAF NBC kit is produced by ML Lifeguard Equipment Ltd.

UNITED KINGDOM - MK.V

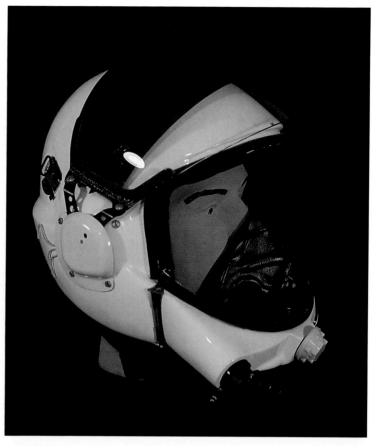

Used in limited numbers by Vulcan bomber crews, in the 1970s, the MK.V is unique with its hinged face piece/oxygen mask combination. Produced by Helmets Ltd. Although it appears as a high-altitude pressure helmet, the MK.V in fact is a conventional helmet.

Yellow knob is depressed and turned to properly tension oxygen mask on face. Oxygen intake is on left, exhaust on right side of hinged face piece. The integrated oxygen mask/protective chin piece hinge open to the right for easy donning.

Single visor has unusual two color design with lower portion clear to read instruments. Shell is fiberglass. Interior is similar to MK.IV.

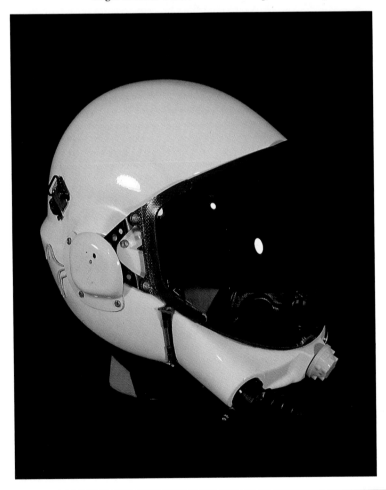

UNITED KINGDOM - ALPHA SERIES

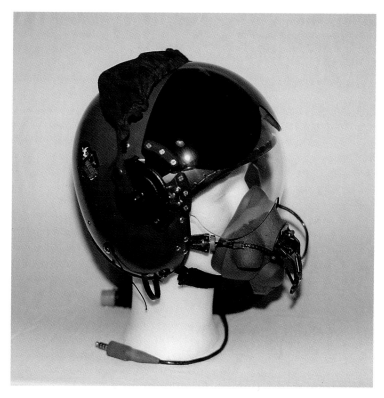

Advanced Lightweight Protective Helmet, Aviation (ALPHA) helmets are the latest models produced by Helmets Ltd. As with the MK.IV helmets, the 1990s ALPHAs are designed for upgrades and interchangability. This is a lightweight Series 700 for use in fighter aircraft. It has a dual visor and uses a "P" mask.

ABOVE AND BELOW: RAF ALPHA with MEL Aviation Ltd. "P" mask. 1986 model visor operated from both sides. RAF Type 10 ALPHA with "P" mask, laser-protective visors which operate from both sides and are infinite locking type.

NVG's mounted on helicopter model with protective visor cover.

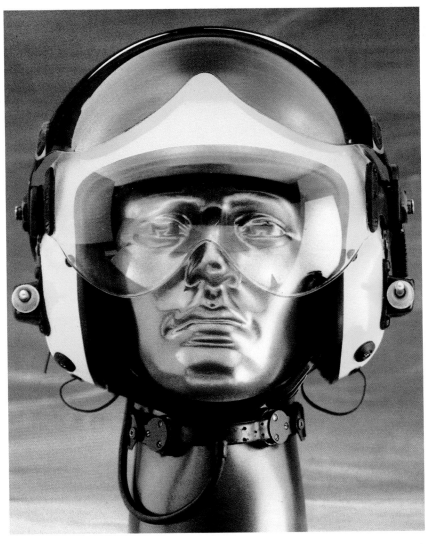

ALPHA Rescuer model uses hands-free communications for search and rescue crews using advanced throat microphones and water-proof radio. Dual visor is standard.

UNITED KINGDOM - MK.10

Limited issue SAS special operations helicopter model. Dual visors, boom microphone, and custom DPM night camouflage paint. Oxygen improves night vision, hence a U.S.-made Gentex MBU-12/P oxygen mask is standard. Super lightweight carbon fiber shell by Helmets Ltd.

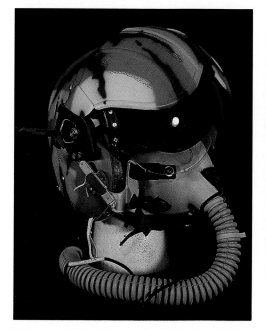

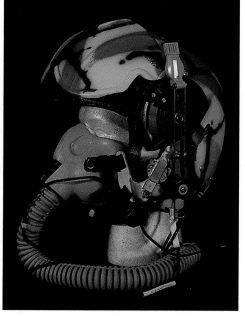

RAF model MK.10B uses "P" mask as used by 5 Sqn. flying Tornado F-3's in 1992. LO-PRO visor system is laser protective.

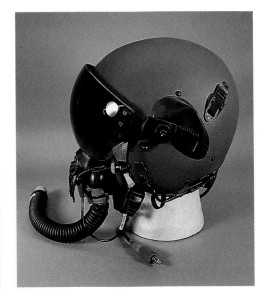

UNITED KINGDOM - G.Q. PARACHUTE CO. LTD. TEST FLYING HELMET TYPE D

Late 1950s model, used in Lightning and Canberra combat aircraft. Partial pressure design is similar in function to the U.S. K-1/MA-2.

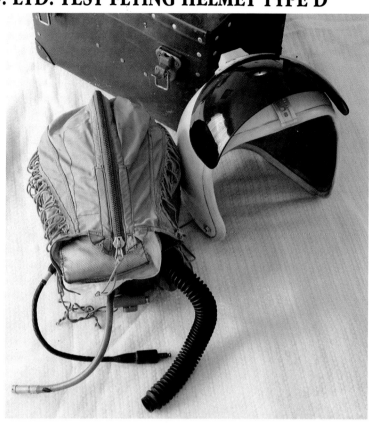

Two part design uses protective fiberglass shell similar to the Mk.1. Issue storage box is also shown.

Dark visor is same as Mk.1.

Microphone switch, communications connector, emergency oxygen hose enter on left side.

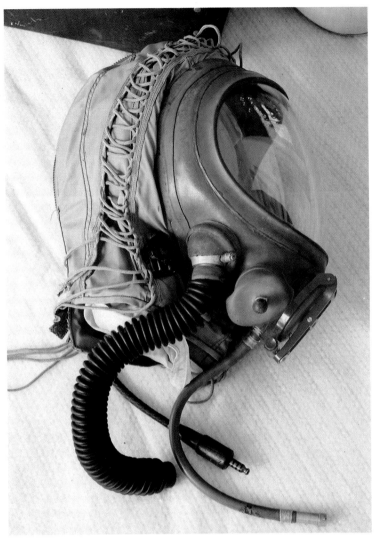

Oxygen hose and exhaust valve mounted on right. Visor is not removable.

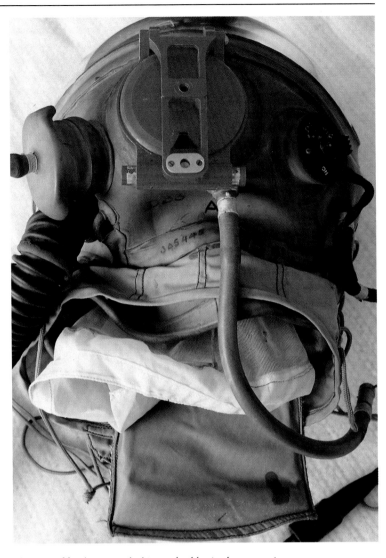

Close-up of food port, neck skirt, and rubberized construction.

UNITED KINGDOM - BAXTER, WOODHOUSE & TAYLOR LTD. TYPE "A"

Shown with issue storage box, bag, visor cover, and documentation.

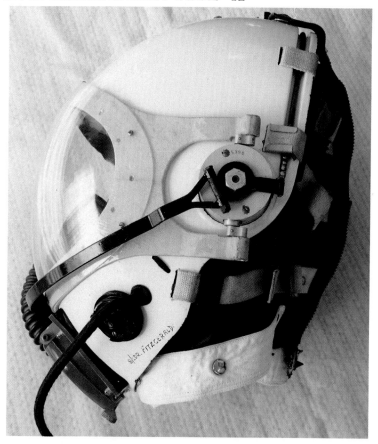

A complex mechanical visor sealing system is used. The spring-loaded tension bar activates vertical cams to seal the visor.

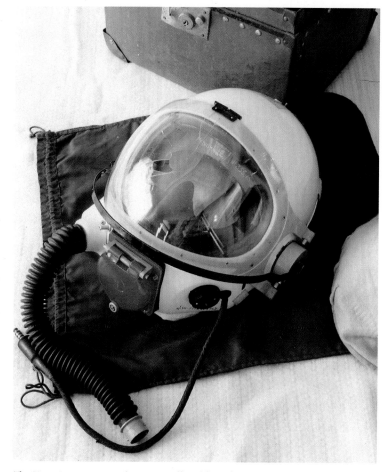

The Type A uses no tinted visor. It suffered from fogging problems.

Face seal is rubber. Visor ventilation openings are seen along the top edge.

UNITED KINGDOM - BAXTER, WOODHOUSE & TAYLOR Ltd. TYPE "E"

Baxter, Woodhouse & Taylor Ltd. Type "E." Taylor Type "E" high-altitude helmet was used by Lightning pilots in the 1960s. The red opening is the mouth port for drinking or vomiting, depending upon the pilot's physical state. The visor is mechanically sealed..

CHAPTER 3: FOREIGN FLIGHT HELMETS - UNITED KINGDOM

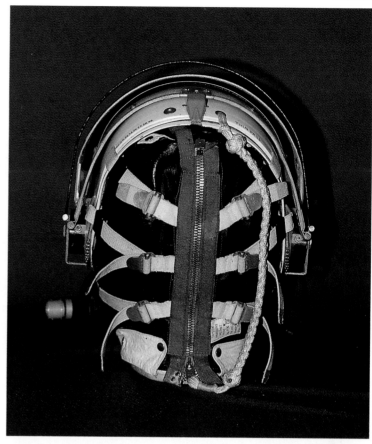

Straps adjust to size.

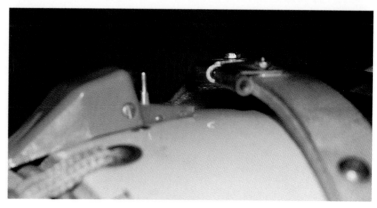

RIGHT: The visor pressure bar is tensioned by spring that runs the length of the crown.

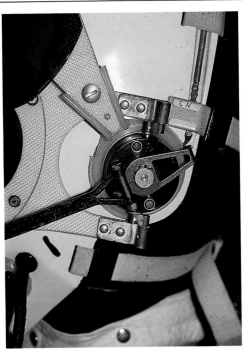

Type "CXP" Lightning test helmet was produced in 1969.

UNITED KINGDOM - MODEL 12P TYPE 4B

Produced by M.L. Aviation Co., LTD., the Type 4B is a high-altitude helmet of the late 1960s. The Type 4B was used by the Civil Aviation Authority Test Pilot for the Concorde SST. Similar in basic design to the Type "E", the red port serves the same purpose.

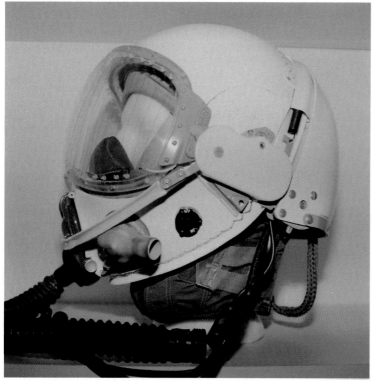
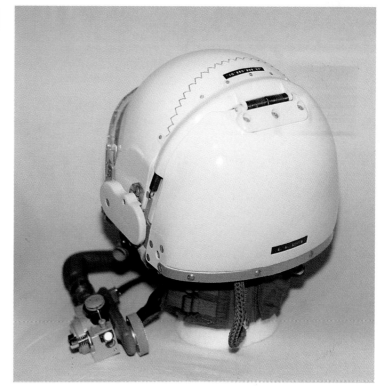

Visor is electrically heated to prevent fogging and is mechanically sealed.

JET AGE FLIGHT HELMETS

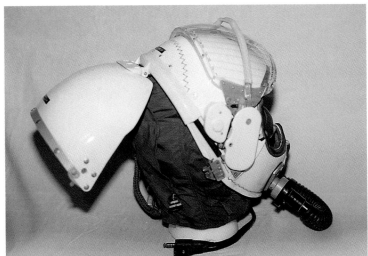

LEFT: The fiberglass shell hinges at the top.

BELOW: Interior is well padded and insulated.

BELOW LEFT: The Model 12P Type 4B modified for use by the Swedish Air Force.

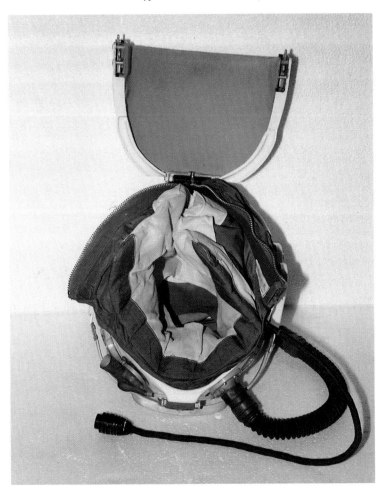

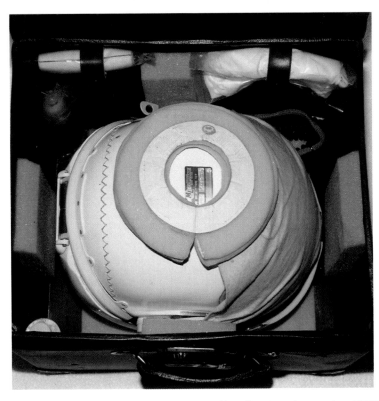

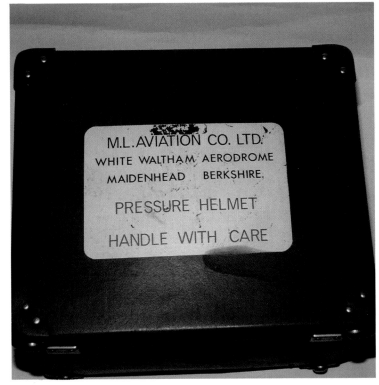

Type 4B in issued storage box. RIGHT: Label on top of issued storage box.

CHAPTER 3: FOREIGN FLIGHT HELMETS - UNITED KINGDOM

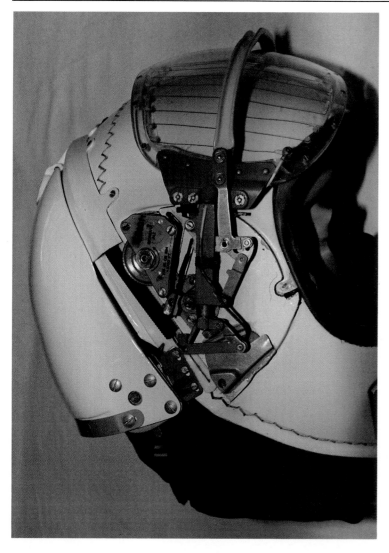

Complex right side visor mechanism is shown.

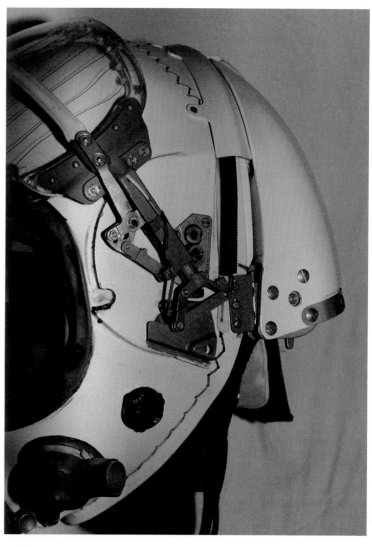

Left side visor mechanism.

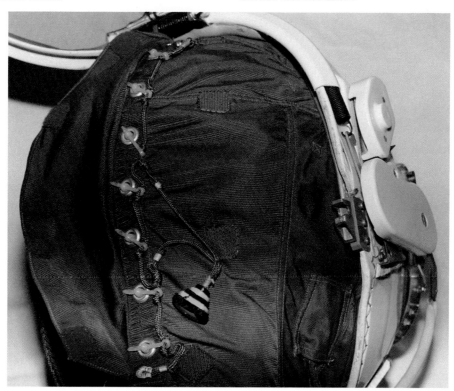

Pressure hood has an emergency escape system using nylon retention pins all connected to the yellow/black knob.

CHAPTER FOUR

SPACE HELMETS

RUSSIA/SOVIET UNION - SK-1

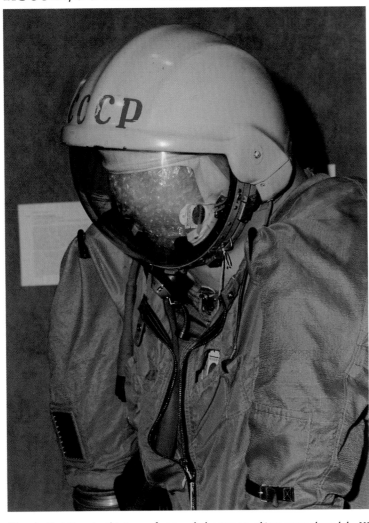 

Worn by Yuri Gagarin, this is very first type helmet to travel into space aboard the VOSTOK spacecraft in 1961. Produced by the Zvesda Works, SK-1 is the entire suit with the helmet having no separate designation.

CHAPTER 4: SPACE HELMETS - RUSSIA/SOVIET UNION

The full-pressure helmet was designed to withstand a sudden decompression of the spacecraft, but also for a high-altitude ejection, which is how Cosmonaut Gagarin returned.

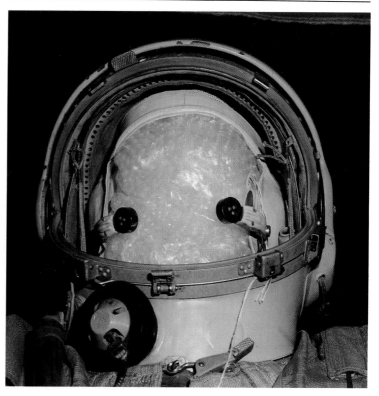

The helmet is rigidly attached to the suit but is large enough that the cosmonaut has room to turn his head within. This fixed-helmet concept reduces neck strain while cosmonaut is enduring long periods of high G's. The sound concept has been adopted in three U.S. space projects: the X-20 Dyna-Soar, Apollo, and current STS "Space Shuttle."

The aluminum SK-1 is well padded. The Oxygen first enters through the perforated strip around the top of the visor to prevent visor fogging.

RIGHT: Visor is mechanically sealed by a latch system still in use today.

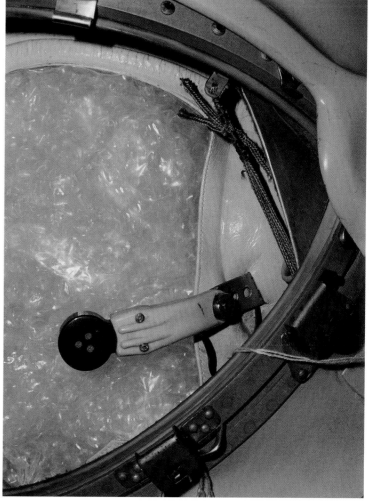

213

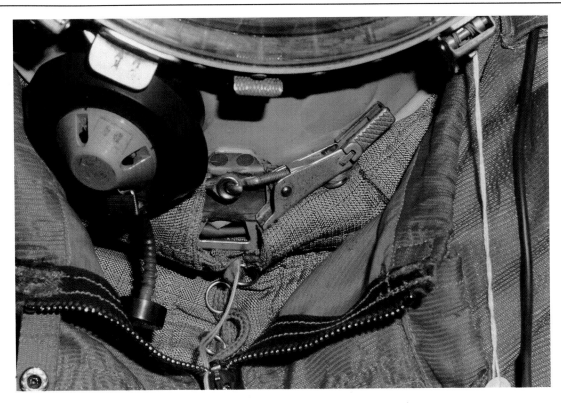
Closeup of suit/helmet securing mechanism and exhaust valve.

Exhaust valve and control cord.

First man in space, Cosmonaut Yuri Gagarin in SK-1 prior to historic flight.

CHAPTER 4: SPACE HELMETS - RUSSIA/SOVIET UNION

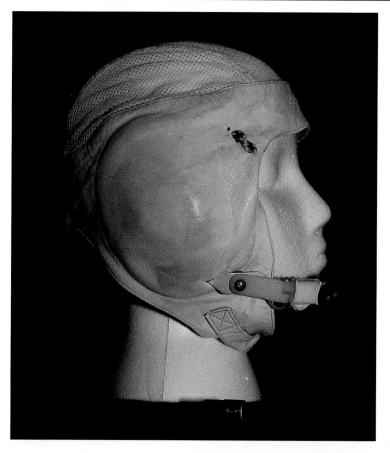

Inner communications helmet is constructed of leather with clothe mesh top.

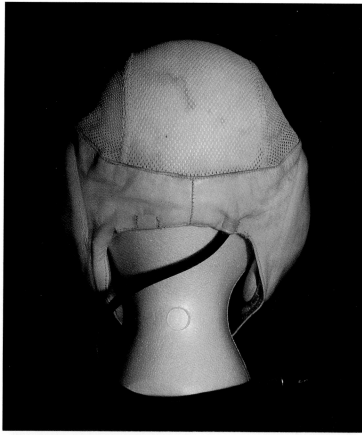

Two microphones were used to provide a backup.

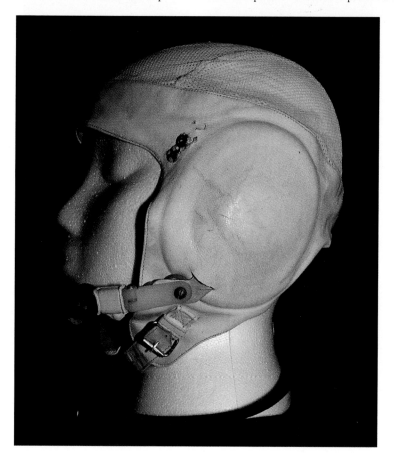

215

RUSSIA/SOVIET UNION - GPSh-3M

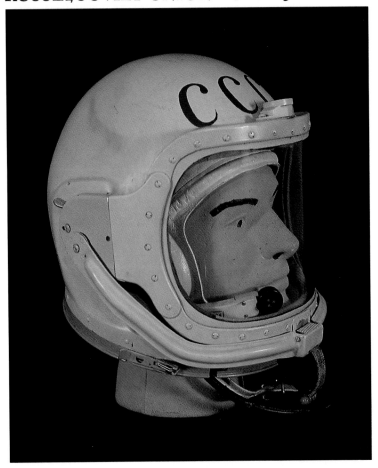

The GPSh-3M was designed by Zvesda to be the Extra Vehicular Activity (EVA) helmet in August, 1963 for the first space walk from the VOSKHOD spacecraft.

An anti-suffocation valve is mounted on the top of the mechanically sealed visor.

Similar in design and appearance to the high-altitude GSh-6, a dark internal visor operates in the same manner. Actuating lever is located behind visor pivot on right side.

Unlike the SK-1, the GPSh-3M is mounted on a rotating neck ring.

Visor is defogged by directing oxygen onto inside first.

CHAPTER 4: SPACE HELMETS - RUSSIA/SOVIET UNION

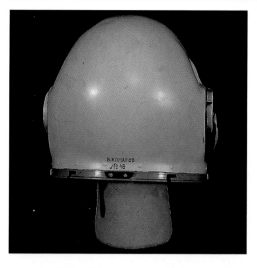  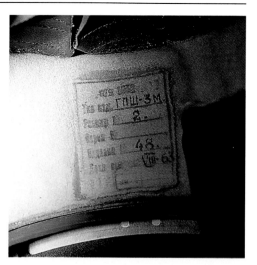

LEFT: The GPSh-3M is constructed of aluminum. CENTER: Padded interior is complimented with adjustable leather supports. RIGHT: As with any space flight equipment, only very small quantities are produced. This model being numbered 48 does not necessarily indicate as many as 48 were ever produced.

Number 48 was used in training by Cosmonaut Vladimir Komarov.

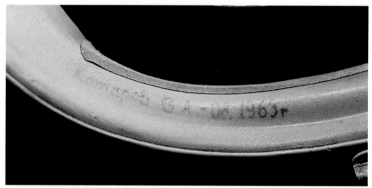

Name and date on inside of visor.

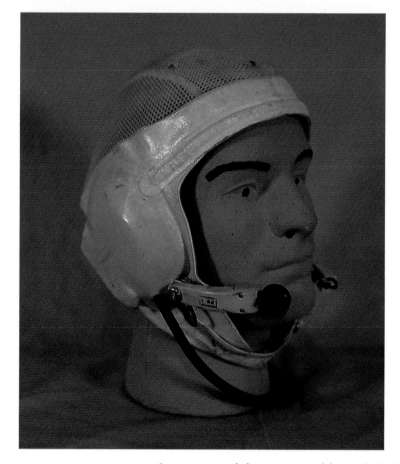 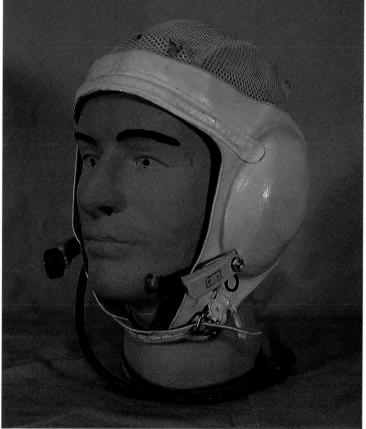

Communications helmet is same model as used with SK-1. This example has broken microphone on left side.

RUSSIA/SOVIET UNION - GPSh-1

The GNSh-1 was a further development of the GPSh-3M helmet. Produced by Zvesda for the "BERKUT" EVA suit in 1965, this was the type used in the very first "space walk" by Cosmonaut Alexei Leonov in March 1965. Another design similar to the high-altitude Gsh-6, the GNSh-1 shares a similar internal dark visor with the actuating lever located on the left front, below the visor pivot.

Used in the VOSKHOD spacecraft, the EVA variant used a gold-tinted visor for protection from radiation.

CHAPTER 4: SPACE HELMETS - RUSSIA/SOVIET UNION

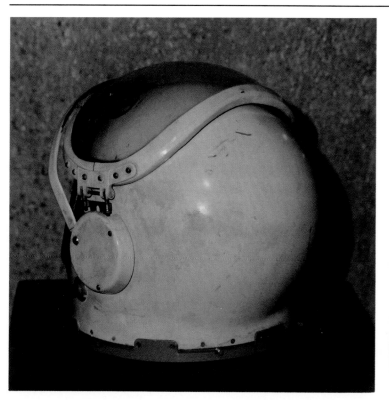
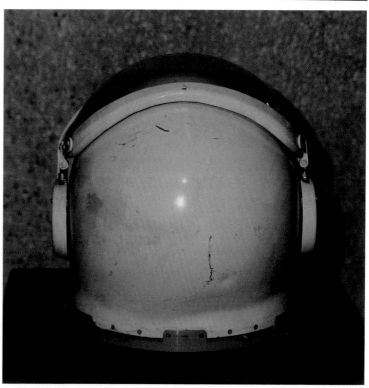

The aluminum shell is attached to the aluminum neck ring by screws.

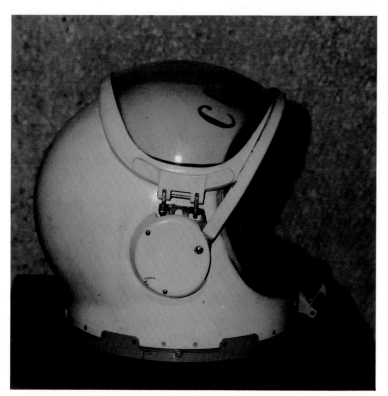

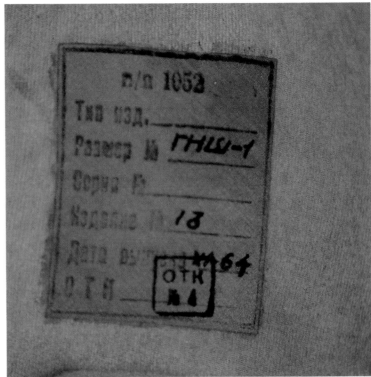

Interior is heavily padded and insulated. A single ventilation/oxygen tube ducts air over the top of helmet and releases it across the visor to prevent fogging.

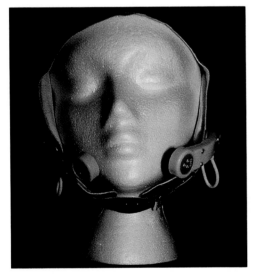
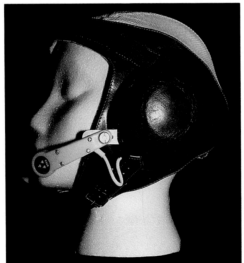
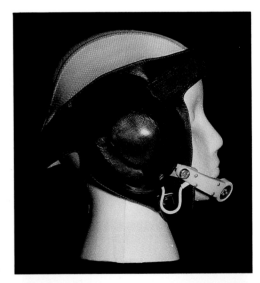

Communications helmet used standard dual microphone arrangement.

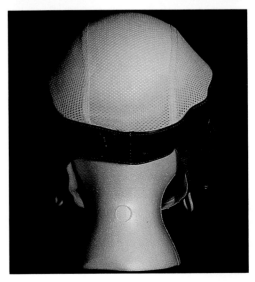

RUSSIA/SOVIET UNION - ShL-2M

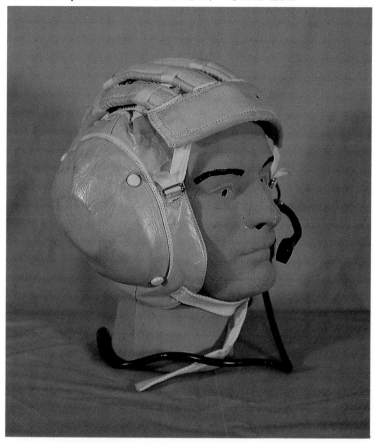
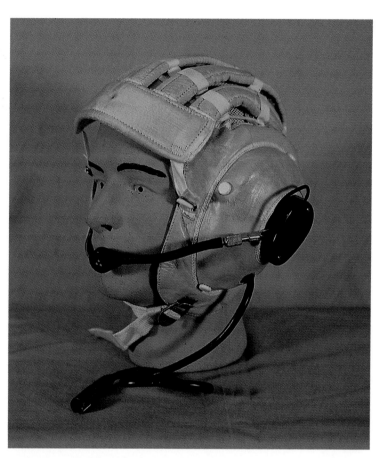

This communications helmet was designed for use in the SALYUT space station. Made of leather with clothe mesh top and removable clothe padding. Produced by Zvesda, helmet is dated 1969.

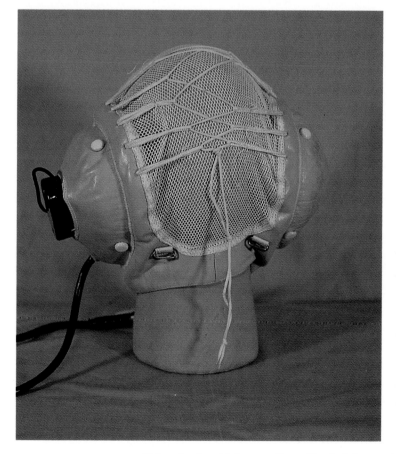
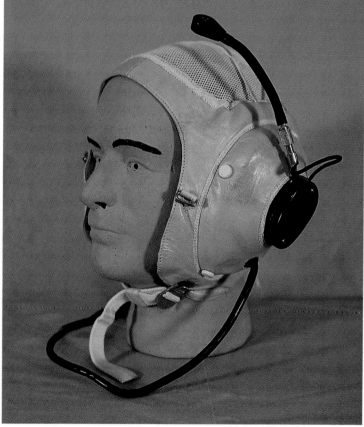

Helmet is adjustable by way of laces. The single boom microphone is flexible and can be pivoted out of the way.

RUSSIA/SOVIET UNION - TNP NZD

Created by Zvesda in 1967, this helmet is part of the one-piece "KRECHET" lunar suit. Secret until only a few years ago, this suit is entered through the rear by way of the life support pack. The helmet is therefore truly part of the whole suit. In 1977 this suit has been refined into the currant "ORLAN" EVA suit used by the "MIR" space station crews. Gold-tinted visor protects from solar radiation.

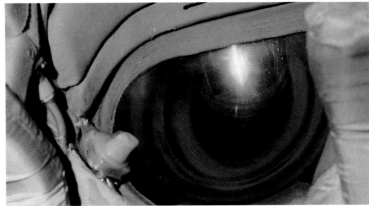

Drinking mouthpiece is mounted on the left. Slots for ventilation are at the top. Suit used GNSh-1 inner communications helmet.

With gold visor stowed.

Label indicates this helmet was produced in 1970, shortly before cancellation of Soviet lunar program.

RUSSIA/SOVIET UNION - "SOKOL"

Zvesda produced "SOKOL I" became the standard Soviet launch and re-entry suit in 1971. The soft pressure helmet is permanently attached to the suit. This system assures a good seal, reduces weight and storage requirements over a hard helmet.

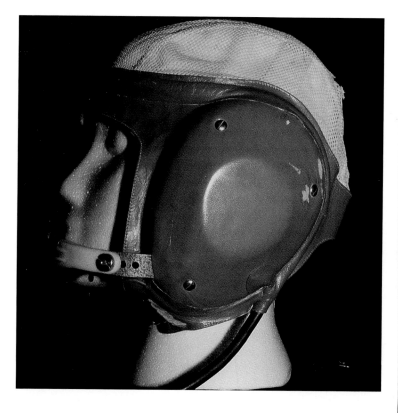

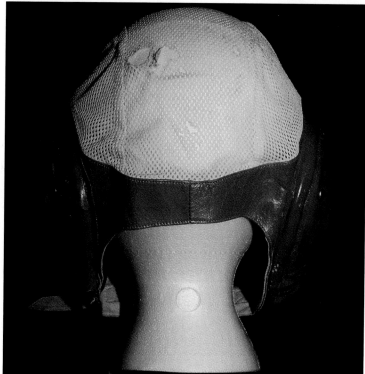

ABOVE AND BELOW: Communications inner helmet is leather with clothe mesh top.

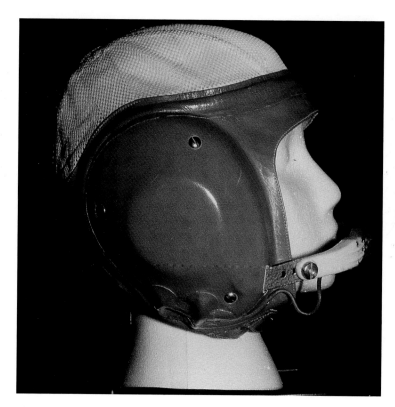

"SOKOL II" suit was introduced by Zvesda in 1982. The helmet area is improved with a larger visor which is double-layered material similar to polycarbonate.

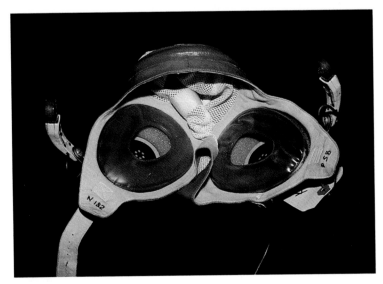

Ear cups are liquid-filled.

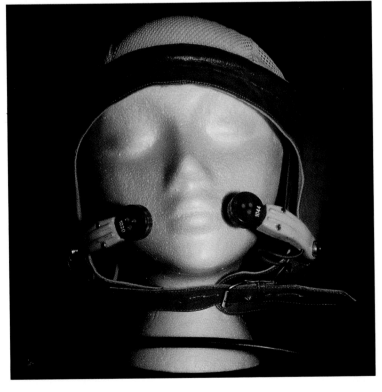

RUSSIA/SOVIET UNION - "SOKOL III"

Introduced in 1990 for use on the "BURAN" space shuttle, the "SOKOL III" has several modifications. Ventilation within the helmet area is improved with larger venting ducts. The camouflage color is reportedly for "Military flights" as the "BURAN" had ejection seats and hence the possibility of a cosmonaut ejecting over hostile terrain.

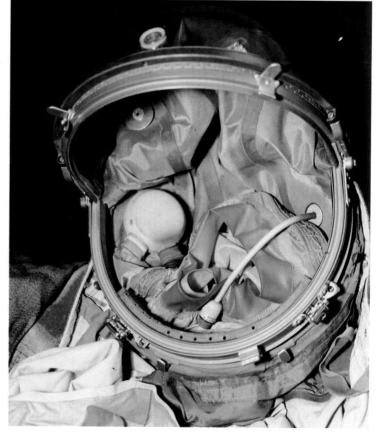

Round exhaust vent can be seen on right. Surgical rubber collar seals oxygen into helmet area. Oxygen is vented into helmet through perforated strip at bottom of opening to prevent visor fogging.

Visor is mechanically sealed using same latching design as used on the 30-year-old SK-1. Exhaust vent on right side.

CHAPTER 4: SPACE HELMETS - RUSSIA/SOVIET UNION

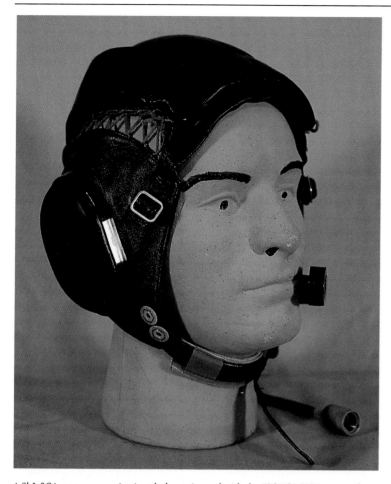
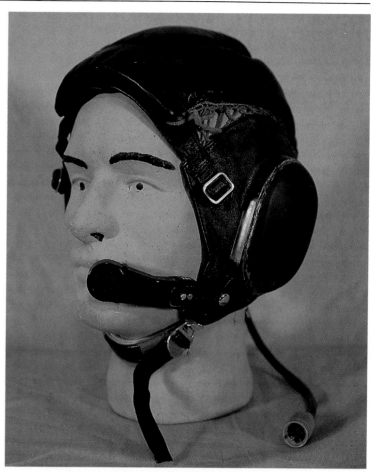

A ShL-9C inner communications helmets is used with the "SOKOL III" It uses only one microphone which is fully adjustable. The helmet is leather with a nylon mesh top. The padded leather crown is removable.

LEFT: Adjustment laces are under padding.

UNITED STATES - PROJECT MERCURY

The helmet worn by the first four Americans in space is based on the USN Mark IV high-altitude model produced by B.F. Goodrich in 1961. A different oxygen system is employed which eliminates the face seal separating the face and the rest of the head as on the Mark IV. The visor seals by way of an inflatable gasket. Leather earphones were production Mark IV type, as were the fiberglass shell and neck attachment ring.

BELOW LEFT: Astronaut Virgil "Gus" Grissom prior to his Mercury/Redstone launch. The dual microphone arrangement can be seen.

BELOW RIGHT: Astronaut Wally Schirra wearing PROJECT MERCURY suit and helmet during training.

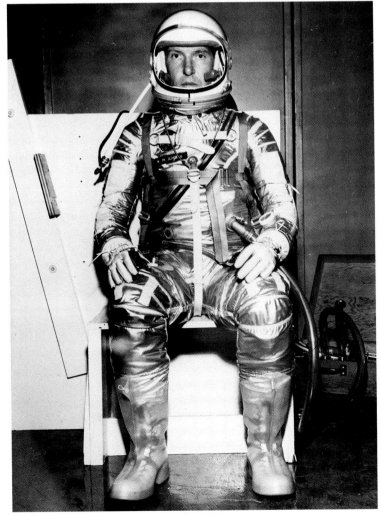

CHAPTER 4: SPACE HELMETS - UNITED STATES

In 1962 a helmet modified with a mechanical visor sealing system was used by Astronaut Gordon Cooper. Contrary to popular thought, the space suits of PROJECT MERCURY were only for emergencies. They were not pressurized during flight. Communications cord contained contacts for more than just the voice; temperature and heartbeat were also monitored as well as perspiration. BELOW LEFT: B.F.Goodrich label denotes helmet simply as "Mechanical Visor Model."

UNITED STATES - X-20 "DYNA-SOAR"

While the USAF X-20 "DYNA-SOAR" program was canceled before completion, the 1962 project resulted in a notable helmet design. As with the successful Soviet SK-1 helmet, the large helmet was designed to permit all necessary head movement from within. This would reduce neck strain during launch. The X-20 was to be a one-man, reusable spacecraft, not unlike the current "Space Shuttle" in concept. The suit and helmet design contract was awarded to the David Clark Co, Inc. This is the first design. The shell is fiberglass with metal framing for added support.

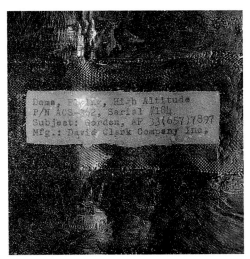

Huge visor is spring-loaded towards the closed position. Lever on lower left adjusts the sizing panel inside the rear of the helmet.

CHAPTER 4: SPACE HELMETS - UNITED STATES

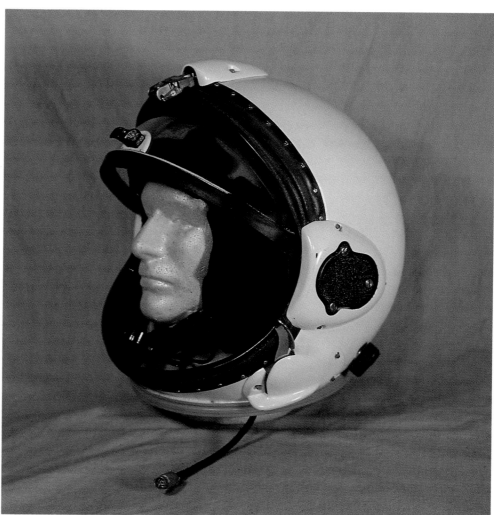

The second type helmet is more wind-resistant, and lighter. This was important as the X-20 was to have an ejection seat. Anti-suffocation valve can be seen on lower right side.

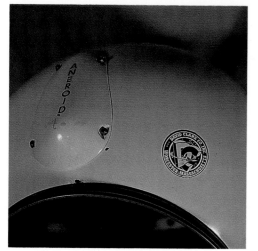

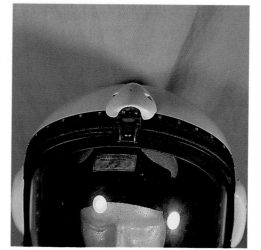

Universal sizing panel is perforated for weight and ventilation purposes.

231

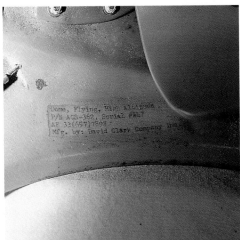

Parachute harness testing of X-20 suit.

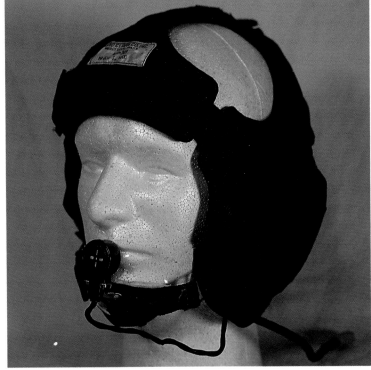

Communications helmet is padded and has microphone mounted uniquely on the chin cup. Helmet is a universal fit with velcro adjustments.

CHAPTER 4: SPACE HELMETS - UNITED STATES

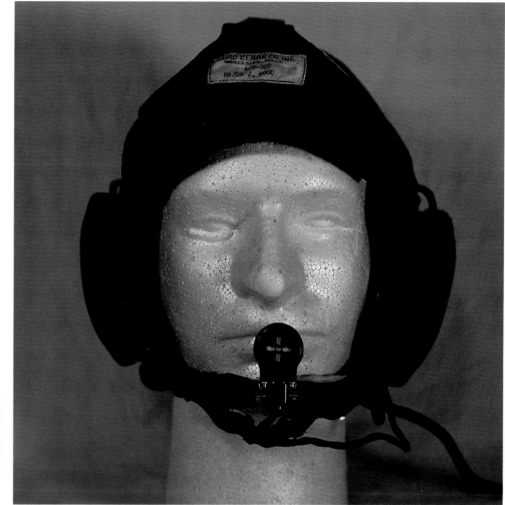

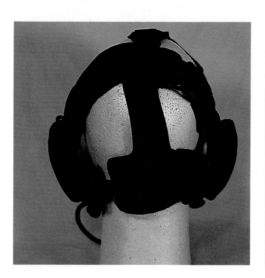

David Clark Co., Inc. logo appeared on their products only on a very limited basis.

UNITED STATES - GH-3C-2

The GH-3C-2 was developed by the David Clark Co., Inc. for PROJECT GEMINI in 1964. It is a fiberglass helmet with a mechanical visor seal system, dual microphones, and standard leather earphones.

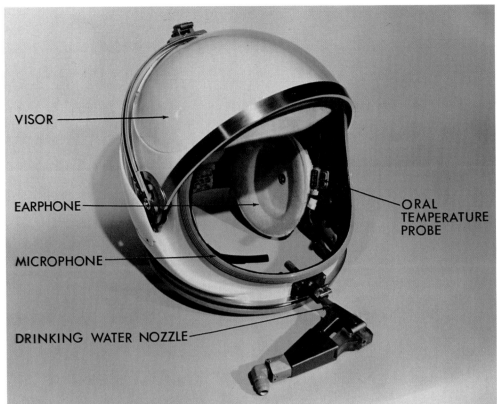

Astronaut John Young during training with GH-3C-2.

CHAPTER 4: SPACE HELMETS - UNITED STATES

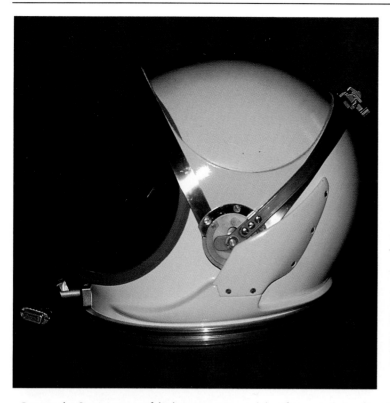

Because the Gemini spacecraft had ejection seats, and therefore, parachutes, fiberglass guards are installed behind the visor pivot to prevent snagging of shroud lines.

A custom-fit liner is leather-covered and has venting holes in two locations. Ducting for these vents can be seen on each side of the neck ring.

Bio-feedback connector is velcroed to the left earphone.

Later type visor latch and food port.

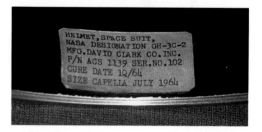

Due to custom fitting, most helmets are size named to the individual, whether an astronaut or test engineer.

Communications connector includes bio-feedback lines as well.

235

UNITED STATES - GH-4C-5

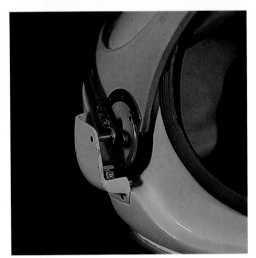

Modified from the GH-3C-2, the GH-4C-5 was designed as NASA's first EVA helmet and was worn in 1965 by astronaut Ed White, the first American to walk in space. Used as the standard helmet on several Gemini missions, the tinted EVA visors were simply omitted. The primary modification is the visor mounting axis is extended, along with the shroud guards, for the mounting of the EVA visors.

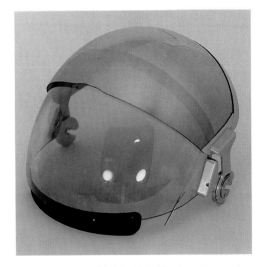

Two visors were added for EVA. The inner Lexan visor has additional UV filtering and micrometeoroid protection while the outer polycarbonate visor is gold tinted for shielding the direct radiation of the sun.

RIGHT: The EVA visors could be positioned in a number of stepped positions as indicated by the notches below the pivot.

Helmet label indicates this model is custom fit for astronaut Frank Borman.

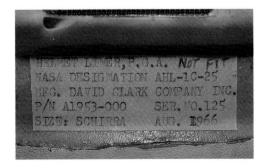

Label of custom fit head padding, indicates fit for astronaut Wally Schirra.

UNITED STATES - G-5C

Designed in 1965 by the David Clark Co., Inc. exclusively for the Gemini VII mission, the helmet of the G-5C suit is the only soft full-pressure helmet ever used in the U.S. space program. Interior space within the Gemini spacecraft was so limited, an astronaut had no room to even remove his helmet, nor was there any place in which to store the item. Hence the need for a soft, pliable helmet which could be easily removed and stored was essential for astronauts Frank Borman and Jim Lovell who spent a record 14-days aboard the cramped spacecraft. A lightweight communications helmet was worn underneath.

UNITED STATES - A-1C

Produced by David Clark Co.,Inc. as the first helmet for PROJECT APOLLO in 1967, the A-1C helmet is a GH-3C-2 modified with a large protective visor shield. Unlike the tiny Gemini spacecraft in which astronauts were fixed in their seats, the Apollo spacecraft afforded, and required, moving about where bumping, scratching, and banging a raised visor would have been a serious problem. Being donned here by Apollo I astronaut Virgil "Gus" Grissom, this type was never used in space, however, as it was replaced when the entire program was redesigned following the fatal Apollo I fire.

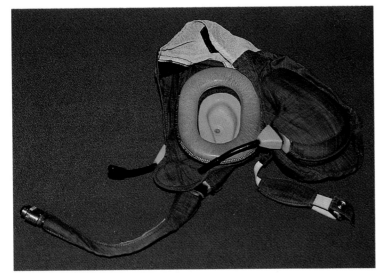

UNITED STATES - A7L

Developed by International Latex Corp., the A7L space suit was adopted by NASA for PROJECT APOLLO in 1968. The A7L helmet is essentially a one-piece, fixed, polycarbonate dome with a head pad in the rear. For use on the Lunar surface, a gold-tinted visor, a white thermal cap, and adjustable eyeshades were added. Shown here being worn by astronaut Neil Armstrong during training.

 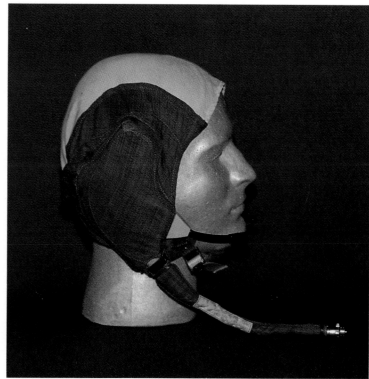

The inner "Communications Carrier Assembly" nicknamed the "Snoopy Cap" used with the A7L suit was designed and produced by David Clark Co., Inc.

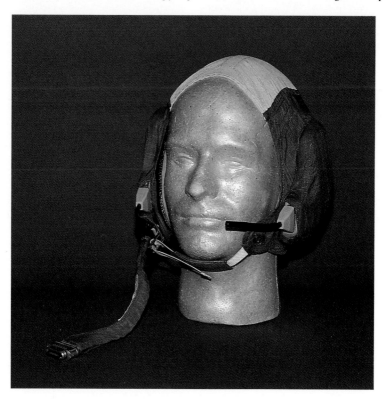

A Nomex mesh was used for comfort and fire protection. Two microphones provide backup.

UNITED STATES - S-1030A

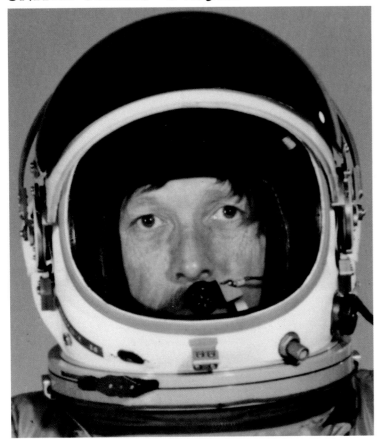

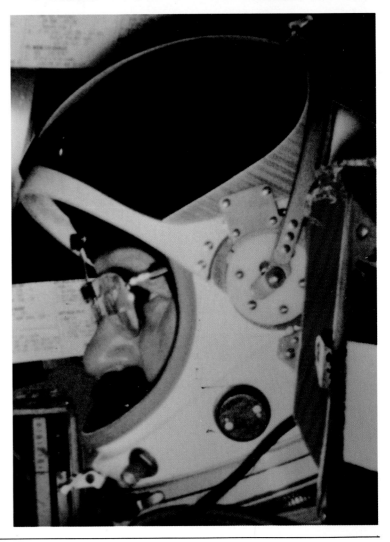

Adopted by NASA in 1979 for use in Space Transportation System (STS) "Space Shuttle" Orbital Flight Test, the S-1030A is a modified S-1030 as used in the SR-71 and is designated as the Emergency Escape Suit (E.E.S.) and later as the Launch and Re-entry suit. Although designated the S-1030A, none of the modifications effect the helmet. It remains unchanged from the S-1030.

RIGHT: Used in the first four space flights, the S-1030A was compatible with SR-71-type ejection seats installed for these first missions. Seen here worn by astronaut Henry Hartsfield Jr.

UNITED STATES - LAUNCH/RE-ENTRY HELMET (LRH)

The fifth STS "Space Shuttle" flight was the first mission considered to be operational. That mission, and the following twenty through the "Challenger" tragedy in 1986, the STS was considered safe enough not to need pressure suits. The requirement for a "shirt sleeve" environment helmet resulted in a helmet heavily based upon the Robert Shaw Co. HGU-20/P "Clamshell." Produced by Gentex using many of the same parts as the HGU-20/P, the LRH has a shell constructed not of fiberglass but of lighter carbon fiber. The oxygen regulator is different and the Gentex LRH has two microphones.

RIGHT: Aboard "Challenger" during re-entry of STS-7 in June, 1983, Astronaut Robert Crippen wears the Gentex LRH. The sizing knob can be seen behind the visor pivot.

UNITED STATES - EMU

For very long duration EVA missions the Extravehicular Mobility Unit (EMU) was adopted for STS in 1982. Produced by Hamilton-Standard, the rigid helmet is constructed of polycarbonate and is essentially a bubble connect to the suit with a neck ring. Inside, ventilation ducts provide fresh oxygen in a defogging flow of air. An inner "Communications Carrier Assembly" almost identical to the PROJECT APOLLO A7L is used and is still produced by David Clark Co., Inc. BELOW: A gold tinted visor and adjustable eyeshades very similar the A7L lunar suit are used. Four small halogen lights are mounted on the top sides of the helmet.

UNITED STATES - S-1032/S-1035

As a result of the STS top-to-bottom re-evaluation program following the "Challenger" tragedy, NASA returned to using a launch and re-entry pressure suit for all Space Shuttle missions. Produced by David Clark Co., Inc., the 1988 model S-1032 and 1992 S-1035 are partial and full pressure suits respectively which use the same helmet. The carbon fiber helmet rotates on its neck mounting ring yet is large enough to permit limited head movement from within. This design was utilized to ease stress on the astronauts' neck during launch when the G-forces, sustained for several minutes, makes head movement very difficult. Astronaut Ken Reightler Jr. during training.

CHAPTER 4: SPACE HELMETS - UNITED STATES

The large polycarbonate visor is mechanically sealed by way of a tension-bar activated cam system much like used on the S-1030 helmet. Dark tinted visor can be positioned as needed. White plastic guards seen behind visor pivots prevent parachute shroud lines from snagging. Inner communications helmet has extra microphone for backup. Astronaut Curtis L. Brown Jr. is hown here on board "Endeavour."

Also from the publisher

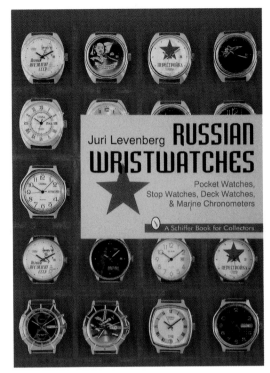

UNITED STATES COMBAT AIRCREW SURVIVAL EQUIPMENT
World War II to the Present - A Reference Guide for Collectors
Michael S. Breuninger

This new book is a detailed study of United States Air Force, Army, Army Air Force, Navy, and Marine Corps aircrew survival equipment. Among the items covered are: survival vests, leggings, and chaps, life preservers, survival (ejection) seat and back pad kits, personal survival kits and first aid kits, and various survival components. Tag and label information is provided for each item.
Size: 8 1/2" x 11" over 170 b/w photographs, drawings
208 pages, soft cover
ISBN: 0-88740-791-9 $29.95

RUSSIAN WRISTWATCHES
Pocket Watches, Stop Watches, Onboard Clock & Chronometers
Juri Levenberg

Collectors have long clamored for a definitive reference and this new book will satisfy even the most avid enthusiast, with photos of over 500 watches manufactured in Russia and the USSR during the second half of this century, and explanations of their styles, workings, and manufacturers. Poljot, Wostok, and Slava wristwatches are covered, along with a sampling of pocket watches, deck watches and marine chrono-meters.
Size: 7" x 10"   505 timepieces in color and b/w photographs   Price Guide
96 pages, soft cover
ISBN: 0-88740-873-7 $19.95

UNITED STATES NAVAL AVIATION PATCHES

A new three volume series covering United States Naval Aviation patches from World War II to the present – each volume contains over 1000 patches in full color:

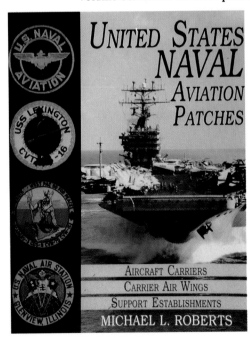

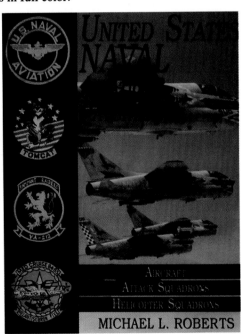

VOLUME I:
AIRCRAFT CARRIERS/CARRIER AIR WINGS
SUPPORT ESTABLISHMENTS
MICHAEL L. ROBERTS
Size: 8 1/2" x 11" over 1000 patches in color, index
160 pages, hard cover
ISBN: 0-88740-753-6 $29.95

VOLUME II:
AIRCRAFT/ATTACK SQUADRONS/HELI SQUADRONS
MICHAEL L. ROBERTS
Size: 8 1/2" x 11" over 1000 patches in color, index
144 pages, hard cover
ISBN: 0-88740-801-X $29.95

VOLUME III:
FIGHTER/FIGHTER ATTACK/RECON SQDNS
MICHAEL L. ROBERTS
Size: 8 1/2" x 11" over 1000 patches in color, index
168 pages, hard cover
ISBN: 0-88740-802-8 $29.95